Lecture Notes Statistics

Edited by D. Brillinger, S. Fienberg, J. Gani, J. Hartigan, and K. Krickeberg

36

Bertil Matérn

Spatial Variation

Second Edition

Springer-Verlag
Berlin Heidelberg New York London Paris Tokyo

Author

Bertil Matérn
Cedervägen 13
S-18245 Enebyberg, Sweden

First edition published by Meddelanden från Statens
Skogsforskningsinstitut, Band 49, No. 5, 1960

AMS Subject Classification (1980): 62D05, 62E99

ISBN 3-540-96365-0 Springer-Verlag Berlin Heidelberg New York
ISBN 0-387-96365-0 Springer-Verlag New York Berlin Heidelberg

Library of Congress Cataloging in Publication Data. Matérn, Bertil, 1917- Spatial variation. (Lecture notes in statistics; v. 36) Bibliography: p. Includes indexes. 1. Sampling (Statistics) 2. Distribution (Probability theory) I. Title. II. Series: Lecture notes in statistics (Springer-Verlag); v. 36. QA276.6.M38 1986 519.5'2 86-13022

Printed in Germany

Printing and binding: Druckhaus Beltz, Hemsbach/Bergstr.
2147/3140-543210

Preface to the Second Edition

This book was first published in 1960 as No. 5 of Volume 49 of Reports of the Forest Research Institute of Sweden. It was at the same time a doctor's thesis in mathematical statistics at Stockholm University.

In the second edition, a number of misprints and other errors have been corrected. An author index and a subject index have been added. Finally, a new postscript comments on the later development of the subjects treated in the book.

BERTIL MATÉRN

March 1986

Acknowledgements

The completion of this thesis was facilitated through the generous assistance of several persons and institutions.

I would wish to express my sincere gratitude to my teacher, Professor Harald Cramér, now chancellor of the Swedish universities, for his valuable help and encouragement. Sincere thanks are also offered to Professor Ulf Grenander for kindly reading the first version of the manuscript and giving valuable advice.

The thesis has been prepared during two widely separated periods. A preliminary draft of Ch. 2 was written in 1948, whereas the remaining parts were completed in 1959—1960. The work originates from problems which I discussed in a publication in 1947. The problems were assigned to me by Professor Manfred Näslund, former head of the Swedish Forest Research Institute, now governor of the province Norrbotten. It is a pleasure to acknowledge my gratefulness to Professor Näslund for his unremitting encouragement and interest in my work. Heartiest thanks are also extended to Professor Erik Hagberg, the present head of the Institute, for many interesting discussions of forest survey problems and for his valuable support.

To *the Faculty of Mathematics and Natural Sciences* of the University of Stockholm I am indebted for a subdoctorate scholarship held during the first six months of 1948. Special thanks are due to *the Board of Computing Machinery* in Stockholm for granting free machine-time on the computer Facit EDB.

Sincere thanks are also offered to Mr. Åke Wiksten, M.F., for his linguistic revision, Mr. Olle Persson, Civ. Eng., for reading the paper in proof and making valuable suggestions, Miss Greta Nilsson and Miss Maud Enström for performing most of the manual calculations, and Mrs. Anne-Liese Neuschel for drawing the figures.

Finally, I am much indebted to many colleagues at the Forest Research Institute for inspiring discussions of topics dealt with in the thesis.

Stockholm, 16 June 1960.

Contents

Chapter I. Introduction

1.1. Preliminary remarks

Patterns of spatial variation are often so complex that only a statistical description can be attempted. A few examples may suffice to support this statement:

the spatial arrangement of microscopic particles suspended in a liquid or in air,

the distribution of galaxies in space,

the pattern of various rock formations on a geologic map,

the spatial distribution of plants or animals in the field, of trees in a forest,

the variation of the tensile strength of a piece of metal,

the microscopic pattern of the surface of a manufactured product (photographic film, sheets of veneer, paper, metal, etc.).

In an earlier paper (1947), the author used the term *topographic variation* to denote the local arrangement of such factors as fertility, vegetation, geologic and climatic occurrences. This term has been adopted by some other writers (B. Ghosh 1949, Whittle 1954). It will be used also in the present paper to denote this somewhat loosely indicated subclass of the spatial variation. Some of the other types of spatial variation referred to above, have been considered e. g. by Neyman & Scott (1958), Fox (1958), Zubrzycki (1957, 58), Savelli (1957), and Husu (1957).

The development of concepts and terms for a description of the properties of the spatial variation may be of value in several situations. In the first place it may be helpful when investigating the underlying mechanism. It may provide a means of specifying certain properties of a manufactured product which are of technical or economic significance. Furthermore, from the statistician's point of view, it is important to have a good knowledge of the spatial variation in a region where a sample survey, or a field experiment is to be arranged.

In the author's paper (1947), already mentioned, a model of the topographic variation was used as the basis for a discussion of the problem of estimating (from the data of the survey) the sampling error of a systematic sample. In the following years the author has encountered various similar problems in sampling, especially in forest surveys. Problems of spatial variation have appeared also in connection with designs of field experiments.

This paper reviews some basic theory of the spatial variation and describes a number of applications to problems in sampling. Applications to questions in field experimentation will be dealt with in a later paper.

The literature of mathematical statistics contains a large number of investigations of phenomena varying in *time*. The interest in time-series is the background of the development of the theory of *stochastic processes* which is at present one of the major fields of probability.

The theory of stochastic processes can be extended to cover also models of spatial variation. This has already been done in several contexts. In some expositions the "index set" of the processes is defined in a general manner. It can then be specialized e. g. as the three-dimensional space, also a time axis may be included. Some authors designate this extension of the stochastic processes as *stochastic* (or *random*) *fields* (e. g. Yaglom 1957). In agreement with Kendall & Buckland (1957) the term stochastic process will here be applied also in the multi-dimensional case (cf. also Bartlett 1955, p. 13).

For the present purpose only a special class of stochastic processes will be utilized. One restriction is that only *static models* will be considered. Of course the dynamic aspect is of importance in describing many phenomena showing spatial variation (cf. § 3.6 below). Cases where the development in time is essential (e. g. turbulence, Brownian motion) are therefore beyond the scope of this paper.

Another limitation is that only *spatially homogeneous* processes will be considered. In analogy with the terminology used in the theory of time-series, such processes will here be called *stationary* (not to be confused with the term "static" in the preceding paragraph). To obtain reasonably realistic models of actual phenomena it may often be necessary to superimpose some inhomogeneous (stochastic or deterministic) "long-distance" component. In the applications the type of description used for this component is often immaterial, see further § 4.1.

This paper should be regarded as an attempt to illustrate the usefulness of the stochastic process approach to statistical questions associated with various types of spatial variation. The field is very broad and interesting extensions seem possible in many directions.

1.2. Survey of the contents

A brief exposition of the mathematics of stationary processes is presented in Chapter 2. The exposition refers to processes in the n-dimensional Euclidean space, R_n. It is restricted to the variance-covariance properties of the processes.

Chapter 3 is devoted to some mechanisms producing sample functions (realizations) of stationary processes in R_n. The object of this study is to give some clues as to the assumptions that are appropriate in schematic models of real phenomena.

Chapter 4 is concerned with the topographic variation. Some empiric data are presented. A discussion of the influence of errors of observation is included. Some particular questions, such as the influence of competition on the distribution of plants, are also touched upon.

Chapter 5 treats the problem of sampling a plane region by a finite number of sample points. Various schemes of selecting the sample points are compared as to their efficiency. The chapter is summarized in § 5.1.

Whereas Chapter 5 deals with one particular problem in some detail, the concluding Chapter 6 is devoted to brief remarks on a number of different questions, most of them associated with forest surveys. Section 6.1 gives a summary of the topics treated in the chapter.

1.3. Notation

The chapters are divided into sections. The formulas are numbered e. g. as (2.3.4) meaning formula 4 of Chapter 2, section 3. When referred to in section 2.3 this formula is quoted as (4).

The following abbreviations are used:

ch.f. characteristic function
cor.f. correlation function
cov.f. covariance function
d.f. (cumulative) distribution function

Vectors and *matrices* are written in ordinary letters .(To avoid confusion, some particular conventions are introduced for Chapters 2 and 3, see § 2.3.) A *transpose* is denoted by the prime mark. A vector $u = (u_1, u_2, \ldots, u_n)$ shall always be understood as a column vector, hence the row vector is always written with a prime, u'. In consequence, the *inner product* of the vectors u and y is written $u'y = y'u$.

An asterisk is used for the *complex conjugate*.

As to general mathematical-statistical concepts, the terminology is chosen in accordance with Cramér (1945).

It should be observed that no distinction is made in the notation between a stochastic process and a realization of such a process.

Chapter 2. Stationary stochastic processes in R_n

2.1. General concepts

In this chapter certain concepts and theorems in the theory of stochastic processes are reviewed, with emphasis on the variance-covariance properties of stationary processes in the n-dimensional Euclidean space, R_n. For details and proofs the reader is referred to one or the other of the following textbooks: Bartlett (1955), Blanc-Lapierre & Fortet (1953), Doob (1953), Grenander & Rosenblatt (1956), Yaglom (1959). Fundamental works such as Khintchine (1934), Wold (1938), Cramér (1940), Karhunen (1947), and Loève (1948) mark the development of the more specific "correlation theory" of stationary processes.

Assume that to every point x of R_n is attached a random variable

$$z(x) = u(x) + i\, v(x) \tag{2.1.1}$$

where u and v are one-dimensional real random variables. Suppose further that $E[|z(x)|^2]$ is finite for every x. The *mean function* of z is

$$m(x) = E[z(x)]$$

whereas the *covariance function* (cov.f.) is defined as

$$c(x, y) = E[z(x)\, z^*(y)] - m(x)\, m^*(y) \tag{2.1.2}$$

Although all applications will deal with real processes, the more general complex form (1) is chosen since it offers notational advantages.

A function $c(x, y)$ is admissible as cov.f. in R_n if, and only if, it is nonnegative definite.

The function

$$r(x, y) = c(x, y) \cdot [c(x, x)\, c(y, y)]^{-1/2}$$

is the *correlation function* (cor.f.) of $z(x)$.

Various examples of covariance functions are found in the following sections. On the basis of such functions others can be constructed by applying the following theorems (cf. Loève 1948, pp. 304 - 5). The class of covariance functions of processes in R_n is here denoted by C.

If $c_1, c_2 \in C$, then $c_1 c_2 \in C$.

Let $\mu(u)$ be a measure in U, and suppose that $c(x, y; u)$ is integrable over the subset V of U for every pair (x, y), and write

$$c(x, y) = \int_V c(x, y; u)\, d\mu(u) \tag{2.1.3}$$

If $c(x, y; u) \in C$ for all $u \in V$, then $c(x, y) \in C$.

If $c_k(x, y) \varepsilon C$ for $k = 1, 2, \ldots,$ and if

$$c(x, y) = \lim_{k \to \infty} c_k(x, y)$$

exists for all pairs (x, y), then $c(x, y) \varepsilon C$.

Occasionally, we shall consider two or more processes simultaneously. Let z_1 and z_2 be two different processes. A *cross covariance function* is defined as

$$c_{12}(x, y) = E[z_1(x)\, z_2^*(y)] - m_1(x)\, m_2^*(y)$$

where m_1 and m_2 are the respective mean functions. Clearly $c_{12}(x, y) = c^*_{21}(y, x)$. In this context the function (2) is more fully designated as an *autocovariance function*. The corresponding nomenclature is used for correlation functions.

A family of processes $z_t(x)$ with t belonging to some index set T, can be considered as a process in the product space (R_n, T). From this representation consistency conditions for cross covariance functions can be deduced, see Cramér (1940).

It may be added that the definitions and theorems presented in this section are valid for processes in a general space.

2.2. Stationary processes

Stationarity is here conceived in the wide sense: A stochastic process is stationary if the mean function is constant and the cov.f. $c(x, y)$ depends on the difference $x - y$ only. We then simply write $c(x - y)$ instead of $c(x, y)$. Similarly the cor.f. is written $r(x - y)$.

In this case the correspondence

$$c(x - y) = c(0)\, r(x - y)$$

exists between the two classes of functions. For this reason the attention is confined to one of them. Choosing to deal with the correlation function we introduce the following symbols: C_n for the class of all functions which can be correlation functions in R_n; C_n' for the subclass of functions which are continuous everywhere except possibly at the origin; C_n'' for the subclass of everywhere continuous functions. In the applications (Ch. 5 – 6) it will be assumed that a cor.f. always belongs to C_n''.

Two stationary processes are *stationarily correlated* if their cross covariance function $c_{12}(x, y)$ depends on the difference $x - y$ only. The two processes are *uncorrelated* if c_{12} vanishes identically.

We shall briefly comment on the two classes C_n' and C_n''. Consider first the following cor.f. belonging to C_n':

$$r_0(u) = \begin{cases} 1 & \text{if } u = 0 \\ 0 & \text{otherwise} \end{cases} \qquad (2.2.1)$$

If $r(u) \varepsilon C_n'$, it can be written

$$r(u) = a\, r_0(u) + b\, r_1(u) \tag{2.2.2}$$

where r_0 is given by (1), $r_1 \varepsilon C_n''$, and a, $b \geq 0$. Proof. Consider a process $z(x)$ with cor.f. $r(u) \varepsilon C_n'$. Write $r(0+)$ for the limiting value of $r(u)$ when $u \to 0$, and assume $r(0+) < r(0) = 1$. To simplify the notation suppose further that the variance, $c(0)$, is 1. Form a sequence $\{z_N(x)\}$ of stochastic processes with

$$z_N(x) = \frac{1}{N} \sum_{j=1}^{N} z(x + d_{j,N}) \tag{2. 2. 3}$$

Assume that for fixed N all $d_{j,N}$ are different and that

$$\operatorname*{Max}_j |d_{j,N}| \to 0 \text{ when } N \to \infty$$

It follows that the corresponding sequence of cov.f.'s converges towards the everywhere continuous function

$$c_1(u) = \begin{cases} r(0+) & u = 0 \\ r(u) & u \neq 0 \end{cases}$$

It is then seen from the closeness properties (cf. 2.1) that c_1 is a cov.f. Thus $r(0+) \geq |r(u)| > 0$ for all u. In the case $r(0+) = 0$, (2) takes the form $r(u) = r_0(u)$. If $r(0+) > 0$, we get the identity (2) by choosing

$$r_1(u) = c_1(u)/r(0+) \qquad b = r(0+) = 1 - a$$

The above proof also indicates that we can decompose $z(x)$ into two components

$$z(x) = z_0(x) + z_1(x) \tag{2.2.4}$$

where z_0 and z_1 are uncorrelated. The "*chaotic component*" z_0 has the cor.f. (1); the "*continuous component*" z_1 has a cor.f. belonging to C_n''. Evidently z_1 is the limit in the mean of the sequence (3).

The following theorems are fundamental in the theory of stationary processes.

If $r \varepsilon C_n$ and is continuous at the origin, then $r \varepsilon C_n''$.

If $r \varepsilon C_n''$, there exists an n-dimensional random variable which has r as characteristic function (ch. f.). The d. f. of this variable is called the *spectral distribution function* of the process. The spectral d. f. of a process thus is a function $F(x)$ where x is a point in R_n. If the corresponding frequency function (probability density) $f(x)$ exists, it will be called the *spectral density* of the process. Conversely, the ch.f. of any n-dimensional random variable belongs to C_n''.

The operation of obtaining a cor.f. from the corresponding spectral d. f. (or vice versa) is *linear*. This property can be utilized to give whole classes of cor.f's when some particular cases are known.

2.3. Isotropic processes

In the sequel *isotropic processes* will often be used as models of the spatial variation. We shall therefore treat such processes in some detail. Further information about these processes and the mathematics involved can be obtained from Bochner (1932, esp. Ch. 9), Schoenberg (1938), Hartman & Wintner (1940), Lord (1954), and Yaglom (1957).

As in the case of stationarity, isotropy is also conceived in the wide sense. Hence, the isotropic process is characterized by the property that the cor.f. is independent of direction. It will be understood that the isotropic process also is stationary.

The notation is simplified by making the following agreements for the rest of this chapter and the following chapter. If a function $H(x)$ with $x \varepsilon R_n$ depends only on the real number $v = |x|$, i. e. the distance of x from the origin, it is simply written as $H(|x|)$ or $H(v)$. Unless explicitly stated otherwise, the letters x, y, and u will be used to denote points in R_n, whereas v, w, and t will denote scalars.

Next, consider an isotropic cor. f. $r(v)$ where v means distance in R_n. The class of such functions will be denoted by D_n. It is a subclass of the class C_n, which was introduced in the preceding section. The corresponding subclasses of C_n' and C_n'' are denoted by D_n' and D_n'', respectively.

Since a cor.f. is Hermitian, all members of D_n are real functions. Since $R_n \subset R_{n+1}$, it is further seen that

$$D_1 \supset D_2 \supset D_3 \supset \dots \qquad (2.3.1)$$

The classes D_1, D'_1 and D_1'' contain all real members of the respective C-classes. D_1'' can also be described as the class containing all real parts of ch.f.'s of one-dimensional random variables. However, for $n > 1$ far more restricted parts of C_n are obtained, as seen from the following theorems.

If $r(v) \varepsilon D_n$, then $r(v) \geqslant -1/n$. Proof: In R_n we can select $n+1$ points so that all mutual distances are equal to a prescribed number v. Let $x_0, x_1, \dots, x_n$ have this property. Then if the mean function of $z(x)$ is put equal to zero,

$$E[|\Sigma z(x_i)|^2] = (n+1)\ c(0)\ [1 + n\ r(v)]$$

This must be non-negative, hence the proposition. If $r \varepsilon D_n'$, the theorem can be sharpened, vide infra.

If $r(v)$ ε D_n with $n>1$, and $r(v_0)=1$ for some $v_0>0$, then $r(v)$ is identically equal to 1. Proof: Suppose $|x-y| = v_0$. To every v in the interval $(0, 2v_0)$ a point u can be found such that $|u-x| = v_0$ and $|u-y| = v$. Since both $z(y)$ and $z(u)$ have correlation 1 with $z(x)$ a perfect correlation must also exist between $z(y)$ and $z(u)$. Hence $r(v) = 1$, and the theorem follows by induction.

In this context it is of interest to note a conjecture of Schoenberg (1938, pp. 822 – 3), which in the terminology used here would mean that the class $D_n - D'_n$ is empty for all $n > 1$.

Now let $r(v)$ belong to D_n'' with $n > 1$. We write

$$r(v) = E[\exp(iu'X)] \tag{2.3.2}$$

where $v = |u|$, and

$$X = (X_1, \ldots, X_n)$$

is an n-dimensional random variable. From (2) is seen that also the distribution of X is isotropic, i. e. unchanged by rotation. It is therefore determined by the d. f. of $|X|$. This d. f. will be denoted $G(w)$ and be called the *radial d. f.* of the process. If the derivative $G'(w)$ exists it will be called the *radial density*. The corresponding ch. f. will be written $g(v)$.

If the first component of u in (2) is v, then

$$r(v) = E[\exp(ivX_1)] = E[\exp(iv|X|Y)] \tag{2.3.3}$$

where Y is one particular coordinate of a point chosen at random (equidistribution) on the surface of the unit sphere in R_n. Y is independent of $|X|$. If the probability density of Y is denoted h_n, it can be shown (see e. g. Wintner 1940) that

$$h_n(w) = \frac{(1 - w^2)^{(n-3)/2}}{B\left(\frac{n-1}{2}, \frac{1}{2}\right)} \tag{2.3.4}$$

with $-1 \leq w \leq 1$. This is a special case of the Beta-distribution, and it has appeared in similar contexts in several papers, e. g. Lhoste (1925), Thompson (1935). The corresponding ch. f. is

$$\Lambda_k(v) = k!\,(2/v)^k J_k(v) \tag{2.3.5}$$

where J_k is the Bessel function of the first kind, and $k = (n-2)/2$. (The notation Λ_k for the above function is found e. g. in Jahnke & Emde 1945, p. 128.)

On the basis of (3) two representations of $r(v)$ are obtained. First (3) can be written in the form

$$r(v) = E[g(vY)]$$

which gives

$$r(v) = \int_{-v}^{v} \frac{h_n(w/v)}{v}\, g(w)\, dw \tag{2.3.6}$$

Next, writing

$$r(v) = E[\Lambda_k(v\,|\,X\,|)]$$

we find

$$r(v) = G(0) + \int_0^\infty \Lambda_k(vw)\, dG(w) \tag{2.3.7}$$

If the radial density G' exists, the following inversion of (7) is obtained

$$G'(w) = \frac{2}{\Gamma(n/2)} \int_0^\infty r(v)\, (vw/2)^{n/2} J_k(vw)\, dv \tag{2.3.8}$$

In this case the spectral density, i. e. the frequency function $f(w)$ of X, may be inserted in (7) and (8) by means of the equation

$$\Gamma(n/2)\, G'(w) = 2\, w^{n-1} \pi^{n/2} f(w) \tag{2.3.9}$$

For the inversion of (6) we first note that $r(v)$ depends only on the real part $\varphi(w)$, say, of the ch. f. $g(w)$. To every $\varphi(w)$ there correspond cor. f.'s $r_2(v)$, $r_3(v), \ldots$, in D_2'', $D_3'', \ldots$, respectively. For $n = 2$ and $n = 3$ the following relations are obtained

$$\varphi(w) = (1/w) \int_0^w \frac{v}{\sqrt{w^2 - v^2}}\, d\,[v r_2(v)] \tag{2.3.10}$$

$$\varphi(w) - \frac{d}{dw}\,[w r_3(w)] \tag{2.3.11}$$

To express $\varphi(w)$ as a functional of a given $r_n(v)$ with $n > 3$, the recurrence relation

$$r_{n-2}(v) = r_n(v) + \frac{v}{n-2}\, \frac{d}{dv}\, r_n(v) \tag{2.3.12}$$

can be used, followed by an application of (10) or (11). For proofs of the above formulas, see Hammersley & Nelder (1955), see also Faure (1957).

Assuming that the derivative $r'(v)$ exists – which is always true for $n > 2$ (Schoenberg 1938, pp. 822 – 3) – we find from (7)

$$-\,n r'(v) = \int_0^\infty v w^2 \Lambda_{n/2}(vw)\, dG(w) \tag{2.3.13}$$

From the above formulas several properties of isotropic cor. f.'s can be found.

First, by means of the asymptotic development of Bessel functions (Watson 1944, Ch. 7) formula (7) gives for $n>1$

$$\lim_{v \to \infty} r(v) = G(0) \tag{2.3.14}$$

If the degenerate case $G(0) = 1$ is neglected, it is seen that

$$[r(v) - G(0)]/[1 - G(0)]$$

constitutes a cor. f. belonging to D_n''. Hence, any $r(v) \varepsilon D_n''$ can be written in the form

$$r(v) = b + (1 - b)\, r_1(v) \tag{2.3.15}$$

with $0 \leq b \leq 1$ and r_1 a continuous cor. f. with the property

$$\lim_{v \to \infty} r_1(v) = 0$$

From (15) and (2.2.2) it is seen that a cor. f. belonging to D_n' can be decomposed into three parts; a component of the form (2.2.1) is added to those appearing in (15).

If $r(v) \varepsilon D_n'$ with $n > 1$, it is further inferred from (6) that $v\, r(v)$ is not only absolutely continuous, but has a continuity modulus, which is independent of v for $n > 2$. When $n > 2$, and $\varepsilon > 0$ (6) gives

$$|(v+\varepsilon)\, r(v+\varepsilon) - v\, r(v)| \leq 2 \int_v^{v+\varepsilon} h_n\left(\frac{w}{v+\varepsilon}\right) dw + 2 \int^{v} \left| h_n\left(\frac{w}{v+\varepsilon}\right) - h_n\left(\frac{w}{v}\right) \right| dw$$

The right hand member equals ε for $n>2$. Thus in this case

$$|(v+\varepsilon)\, r(v+\varepsilon) - v\, r(v)| \leq \varepsilon \tag{2.3.16}$$

Next, if $r(v) \varepsilon D_n'$, it is found from (7) and (2.2.2) that

$$r(v) \geq \operatorname*{Inf}_{v} \Lambda_k(v) \tag{2.3.17}$$

with $k = (n-2)/2$. Tables of Bessel functions give

$n=2 \quad r(v) > -0.403$
$n=3 \quad r(v) > -0.218$
$n=4 \quad r(v) > -0.133$

It should finally be remarked that (6) still gives a cor. f. such that $v\, r(v)$ is continuous when $n \geq 2$, if $g(w)$ is an integrable but otherwise arbitrary cor. f. This gives further support to the above-mentioned conjecture of Schoenberg. If $z_1(t)$ is a stationary process in R_1 with cor. f. $g(t)$, $z(x) = z_1(x'Y)$ with Y chosen independently of z_1 on the surface of the unit n-sphere, is then an example of a process with cor. f. (6).

2.4. Examples of correlation functions

Examples of continuous correlation functions can be obtained from distributions appearing in statistical theory. We then utilize the identity between the class C_n'' of cor. f.'s and the class of characteristic functions of n-dimensional random variables.

When a normal (gaussian) spectral distribution is chosen, it follows

$$\exp(-u'Au) \in C_n'' \tag{2.4.1}$$

if $u'Au$ is a non-negative quadratic form, see Cramér (1945, p. 310). A special case is the isotropic cor. f.

$$\exp(-a^2v^2) \tag{2.4.2}$$

which belongs to D_n'' for every n. The corresponding radial density is seen to be

$$G'(w) = \text{const. } w^{n-1} \exp(-w^2/4a^2) \tag{2.4.3}$$

By (2.1.3)

$$\int_{-\infty}^{\infty} \exp(-a^2v^2)\, dH(a) \in D_n'' \tag{2.4.4}$$

for all n. Here $H(a)$ is an arbitrary one-dimensional d.f. Formula (4) is *the general expression for a cor. f. belonging to every D_n''*, see Schoenberg (1938, pp. 817 ff.) and Hartman & Wintner (1940, p. 763).

Select then for a^2 a "type III distribution" (see Cramér 1945, pp. 126, 249). With $s, b > 0$, it is found that

$$\int_0^{\infty} \exp[-a^2(v^2+b^2)] \frac{b^{2s}}{\Gamma(s)} a^{2s-2}\, da^2 = (1 + v^2/b^2)^{-s} \tag{2.4.5}$$

belongs to every D_n''. To obtain the corresponding spectral density the same transformation can be applied to the spectral density of the cor. f. (2), which is

$$\text{const. } a^{-n} \exp(-w^2/4a^2)$$

Thus the spectral density of (5) is

$$f(w) = \text{const.} \int_0^{\infty} \exp(-w^2/4a^2 - b^2a^2)\, a^{2s-n-2}\, da^2 =$$

$$= \text{const. } w^{s-n/2} K_{s-n/2}(wb) \tag{2.4.6}$$

cf. Ryshik & Gradstein (1957, formula 3.282). Here K is the modified Bessel function of the second kind, see Watson (1944, p. 78). Lord (1954, p. 55) calls this the "right" generalization to n dimensions of the type III distribution. (Lord considers only cases with $s > n/2$.)

For $s > n/2$ (5) is a frequency function in R_n, if multiplied by a suitable constant. Thus its Fourier transform gives a cor. f. in R_n. Hence

$$2\,(bv/2)^{\nu} K_{\nu}(bv)\,/\Gamma(\nu) \in D_n'' \qquad (2.4.7)$$

if b and ν are $\geqslant 0$. For the constant in (7), see Watson (1944, p. 80).

Two special cases of (7) deserve mention. For $\nu = 1/2$

$$\exp(-bv) \in D_n'' \qquad (2.4.8)$$

for all n. The corresponding spectral distribution is (5) with $s=(n+1)/2$. It is the generalization to n dimensions of the Cauchy distribution (Quenouille 1949, Lord 1954, cf. Bochner 1932, p. 189). For $n=1$ (8) is the cor. f. of a Markoff process. If $\nu=1$ is inserted in (7)

$$vb\,K_1(vb) \qquad (2.4.9)$$

is obtained. It has been called the corresponding elementary correlation in R_2 (Whittle 1954).

Applying (2.1.3) to (8) we find that

$$\int_{-\infty}^{\infty} \exp(-|bv|)\,dH(b) \in D_n'' \qquad (2.4.10)$$

for every n. Here H is an arbitrary d. f. According to Bernstein's theorem (see Widder 1941, p. 160) this is the general form of a function completely monotonic in $0 < v < \infty$ and attaining the value 1 at the origin.

If the total mass in the radial distribution is concentrated in the point a, we find from (2.3.7)

$$\Lambda_k(av) \in D_n'' \qquad (2.4.11)$$

where $k=(n-2)/2$ and Λ is the function (2.3.5). The corresponding spectral distribution is the equidistribution on the surface of the sphere $|x|=a$ in R_n.

It can now be shown that all functions of the type (11) of higher order than k also belong to D_n''. Transforming Sonine's first finite integral (Watson 1944, p. 373) we have with $a, k, s > 0$

$$\Lambda_{k+s}(av) = \frac{2}{aB(s,k+1)} \int_0^a \Lambda_k(vw)\,(w/a)^{2k+1}(1-w^2/a^2)^{s-1}dw \qquad (2.4.12)$$

Hence

$$\Lambda_s(av) \in D_n'' \quad if \quad s \geqslant (n-2)/2 \qquad a > 0$$

as is plausible from (11) and (2.3.1).

A special case of (12) is the cor. f.

$$\Lambda_{n/2}(v) \qquad (2.4.13)$$

which has the equidistribution in the interior of the unit n-sphere as spectral distribution. For other connections with n-dimensional geometry see Wintner (1940) and Hartman & Wintner (1940).

With the exception of the first example, only isotropic cor. f.'s have been dealt with. Now the isotropic cor. f. (2) is a special case of the more general function (1). Likewise, every isotropic $r(v)$ can be regarded as a particular case of a cor. f.

$$r\left(\sqrt{u'Au}\right) \tag{2.4.14}$$

Thus a class of correlation functions is obtained where the "iso-correlation surfaces" are ellipsoidal and not necessarily spherical as in the case of an isotropic cor. f.

The following is a more general way of obtaining a family of cor. f.'s from a given function. Let $z_1(y)$ be a stationary process in R_m with cor. f. r_1. Further, let B be a real matrix of order $m \cdot n$. Define with $x \varepsilon R_n$

$$z(x) = z_1(Bx) \tag{2.4.15}$$

Then $z(x)$ is stationary with cor. f.

$$r(u) = r_1(Bu) \tag{2.4.16}$$

If z_1 has the isotropic cor. f. $r_1(v)$

$$r(u) = r_1\left(\sqrt{u'B'Bu}\right)$$

i. e. the "elliptic" case (14).

Of course, various other operations carried out on a stationary process $z_1(y)$ with cor. f. r_1 can give new stationary processes with cor. f.'s that have more or less complicated relations to r_1. Examples:

$$z_2(y) = f[z_1(y)]$$

$$z_3(y) = \int_{R_n} z_1(y-x)\, g(x)\, dx$$

where f and g are non-stochastic complex-valued functions. See further Ch. 3.

Still other types of cor. f.'s in R_n are obtained by forming the product of cor. f.'s in R_m and R_k with $m+k=n$.

2.5. Integration of stationary processes

Integrals of the type

$$\int_{S_i} \varphi(x)\, z(x)\, dx \tag{2.5.1}$$

where $\varphi(x)$ is continuous over the finite set S_i in R_n, and $z(x)$ is a stationary process with cor. f. $\varepsilon\, C_n''$, can be defined as limits in the mean of finite linear forms in stochastic variables $\{z(x_r)\}$ in analogy with the Riemann definition of an integral, see Cramér (1940). Cramér deals with processes in R_1. The extension to R_n is straightforward. The integration can be carried out under somewhat milder restrictions on φ and z, e. g. for processes with a cor. f. belonging to C_n'.

A more general concept of integration is outlined in Karhunen (1947). Since we shall mainly be concerned with the first two moments of the integrals, there is no need to enter into further discussions of the concept of integration. The integration of a stationary process can be regarded as a special case of a general linear operation (or "filtering") carried out on the process; for a rigorous treatment along these lines, see Doob (1953).

Let $z(x)$ have the constant mean m and the covariance function $\sigma^2 r(u)$ with $r\,\varepsilon\, C_n'$. Then z is integrable in Cramér's sense. Denote the integral (1) by I_i. The first two moments of the integrals are given by

$$E\,(I_i) = m \int_{S_i} \varphi\,(x)\, dx \qquad (2.\,5.\,2)$$

$$\text{Cov.}\,(I_i, I_j^*) = \sigma^2 \int_{S_i} \int_{S_j} r\,(x-y)\, \varphi\,(x)\, \varphi^*\,(y)\, dx\, dy \qquad (2.\,5.\,3)$$

From the discussion leading to (2.2.4) it is clear that r can be replaced in (3) by its continuous component, br_1 of (2.2.2). We thus have

$$E\,(I_i) = \left\{E\,[z_0\,(x)] + E\,[z_1\,(x)]\right\} \int_{S_i} \varphi\,(x)\, dx$$

$$\text{Cov.}\,(I_i, I_j) = \sigma_1^2 \int_{S_i} \int_{S_j} r_1\,(x-y)\, \varphi\,(x)\, \varphi^*\,(y)\, dx\, dy$$

where $\sigma_1^2 r_1$ is the cov. f. of the continuous component, z_1. Thus the presence of the chaotic component, z_0, introduces no serious complications in the formulas. When finite sums are considered instead of integrals (vide infra), the variance of z_0 enters into the formulas in a trivial way. Although such a chaotic component is needed in some applications, it can be disregarded for the moment. Consequently, it is assumed for the rest of this section that z has a cor. f. $\varepsilon\, C_n''$.

The applications in the following chapters mostly refer to the case $\varphi = 1$. It is then convenient to deal with *average values*, which will be denoted in the following way

$$z\,(S_i) = \int_{S_i} z\,(x)\, dx / \mu\,(S_i) \qquad (2.\,5.\,4)$$

where $\mu(S_i)$ is the volume of S_i. Averages over sets of infinite measure might also be defined by a passage to the limit, but such averages are uninteresting from the point of view of the following applications. Their statistical properties are related to the ergodic theorems, cf. Grenander & Rosenblatt (1956, pp. 42 ff.).

The main advantage of using averages instead of integrals is that several averages $\{z(S_i)\}$ can easily be handled simultaneously, even if the sets have varying dimensions from 0 to n.

If S_i belongs to a linear subset of R_n, the integral over S_i and the measure $\mu(S_i)$ in (4) should be understood to refer to this space of lower dimension. Proceeding to a set S consisting of a finite number of points $(x_1, x_2, \ldots, x_N)$ we define

$$z(S) = \frac{1}{N} \Sigma z(x_i)$$

Formulas for mean values, variances and covariances of averages can be expressed in a more compact form than the corresponding formulas for integrals. For the mean value we have

$$E[z(S)] = m$$

whereas the covariance can be written as follows

$$\text{Cov.}\,[z(S_i), z^*(S_j)] = \sigma^2 \int_{R_n} r(u)\, dA_{ij}(u) = \sigma^2 E\,[r(U-V)] \qquad (2.5.5.)$$

Here A_{ij} is the d. f. of the difference $U-V$ between two points chosen independently and at random (equidistribution) in S_i and S_j respectively. If $j=i$, (5) gives an expression for the variance of $z(S_i)$.

If the cor. f. is isotropic, (5) can be replaced by

$$\text{Cov.}\,[z(S_i), z^*(S_j)] = \sigma^2 [B_{ij}(0) + \int_0^\infty r(v)\, dB_{ij}(v)] = \sigma^2 E\,[r(|U-V|)] \qquad (2.5.6)$$

where B_{ij} is the d. f. of the distance $|U-V|$.

The equivalence of (5) and (6) with the expression obtained more directly from (3) follows from theorems on transformations of integrals, see Saks (1937, p. 37) and Simonsen (1947).

Next, we consider formulas expressing the covariances in the spectral and radial distributions. Let $F(u)$ be the spectral d. f. of the process (see § 2.2). By applying Fubini's theorem (Saks 1937, p. 77) on (5) we find

$$\text{Cov.}\,[z(S_i), z^*(S_j)] = \sigma^2 \int_{R_n} \alpha_{ij}(u)\, dF(u) \qquad (2.5.7)$$

where α_{ij} is the ch. f. of the difference $U - V$. Since U and V are independent

$$\alpha_{ij}(u) = \alpha_i(u)\,\alpha_j^*(u) \tag{2.5.8}$$

where $\alpha_i(u)$ is the ch. f. of the rectangular coordinates of a point chosen at random in S_i.

We shall give two examples of ch. f.'s of this type.

If S is the n-dimensional interval

$$c_j - d_j < x_j < c_j + d_j \qquad (j = 1, 2, \ldots, n)$$

the ch. f. is

$$\exp(iu'c) \prod_{j=1}^{n} \frac{\sin(u_j d_j)}{u_j d_j} \tag{2.5.9}$$

Here x_j, c_j, d_j, and u_j denote the j:th components of the vectors x, c, d, and u, respectively.

If S is the ellipsoid

$$(x' - c')\,A\,(x - c) < a^2$$

the ch. f. is

$$\exp(iu'c)\,\Lambda_{n/2}\left(a\sqrt{u'A^{-1}u}\right) \tag{2.5.10}$$

If $r \varepsilon D_n''$ has the radial d. f. $G(w)$ (see 2.3.7),

$$\text{Cov.}\,[z(S_i), z^*(S_j)] = \sigma^2 \left[G(0) + \int_0^\infty \psi_{ij}(w)\,dG(w)\right] \tag{2.5.11}$$

Here ψ_{ij} is the ch. f. of the product of the two independent variables $|U - V|$ and Y, where Y is one particular coordinate of a point chosen at random on the surface of the unit n-sphere, cf. (2.3.4) and (2.3.5). Hence in analogy with (2.3.7)

$$\psi_{ij}(w) = B_{ij}(0) + \int_0^\infty \Lambda_k(vw)\,dB_{ij}(v) \tag{2.5.12}$$

with $k = (n-2)/2$.

An alternative to (11) is

$$\text{Cov.}\,[z(S_i), z^*(S_j)] = E[\beta(|X|Y)].\sigma^2 \tag{2.5.13}$$

where β is the ch. f. of the distance $|U - V|$, and $|X|$ is a random variable with d. f. $G(w)$, whereas Y is one rectangular coordinate of a point chosen at random on the boundary of the unit n-sphere.

In practical computations, one or the other of the above formulas can be found to be convenient. In some cases, when the sets $\{S_i\}$ are irregular in shape, or the integration presents difficulties for other reasons, a formula of the type (6) indicates a way of applying the Monte Carlo method. If $\{U_h, V_h\}$ are N pairs of sample points chosen at random in S_i, and S_j, respectively, the expression

$$\frac{\sigma^2}{N} \sum_h r(|U_h - V_h|)$$

is an unbiased estimate of the covariance between $z(S_i)$ and $z^*(S_j)$.

In the following chapters we shall have to deal with expectations of quadratic forms of the type

$$T = \sum_{i=1}^{N} \sum_{j=1}^{N} c_{ij} [z(S_i) - m] [z^*(S_j) - m^*]$$

From the different covariance-formulas of this section various expressions for $E(T)$ are obtained. E. g. from (7)

$$E(T) = \sigma^2 \int_{R_n} \alpha_T(u) \, dF(u) \qquad (2.5.14)$$

where

$$\alpha_T = \sum c_{ij} \alpha_i \alpha_j^* \qquad (2.5.15)$$

may be called *the characteristic function of the quadratic form T*.

In the isotropic case we may apply (6), e. g. Then

$$E(T) = \sigma^2 [\Delta_T(0) + \int_0^\infty r(v) \, d\Delta_T(v)] \qquad (2.5.16)$$

with

$$\Delta_T = \sum c_{ij} B_{ij} \qquad (2.5.17)$$

Formulas of the type (14) – (16) were used in the author's earlier investigation (Matérn 1947, pp. 32,55). An expression such as (17) was called a *distance-integral*. The derivative of (17), if existing, was called the *distance function* of T.

When the sets $\{S_i\}$ are composed of elements such as rectangles and spheres, it is usually easy to find an explicit expression for the characteristic function (15), see (9) and (10). A deduction of the corresponding distribution functions and frequency functions seems to be more difficult, and the expressions tend to become somewhat unwieldy. However, some special cases will be reviewed for later reference.

These cases all refer to the distribution of the distance between two points chosen at random and independently in a plane convex region. The region is denoted by S, and its area and perimeter by A and P, respectively.

First, let S be a rectangle with the sides A_1 and A_2. The fr. f. was derived by Ghosh (1943), see also Ghosh (1951). It can be written as follows (see Matérn 1947, p. 35):

$$\frac{1}{\sqrt{A}} f\left(v/\sqrt{A},\ \sqrt{A_1/A_2}\right) \tag{2.5.18}$$

where

$$f(w, a) = 2w\,[f_1(w, a) + f_2(wa, a) + f_2(w/a, 1/a)]$$

with

$$f_1(w, a) = \begin{cases} \pi + w^2 - 2w(a + 1/a) & 0 < w < \sqrt{a^2 + a^{-2}} \\ 0 & \text{otherwise} \end{cases}$$

$$f_2(w, a) = \begin{cases} 2\sqrt{w^2 - 1} - 2\arccos(1/w) - a^{-2}(w-1)^2 & 1 < w < \sqrt{1 + a^4} \\ 0 & \text{otherwise} \end{cases}$$

For small v, the frequency function of the distance between two points in S can be written

$$\frac{2\pi v}{A} - \frac{2v^2 P}{A^2} + o(v^2) \tag{2.5.19}$$

This formula is valid for any convex region, cf. Borel & Lagrange (1925, p. 87). The corresponding expression for the d. f. is

$$\frac{\pi v^2}{A} - \frac{2v^3 P}{3A^2} + o(v^3)$$

For a circle with radius R the frequency function is

$$\frac{1}{R} f_C\left(2 \arcsin \frac{v}{2R}\right) \tag{2.5.20}$$

with

$$f_C(w) = \begin{cases} 4 \sin(w/2) \left[1 - \dfrac{w + \sin w}{\pi}\right] & 0 < w < \pi \\ 0 & \text{otherwise} \end{cases}$$

cf. Deltheil (1926, p. 39). The corresponding d. f. for a sphere in R_n is also found in Deltheil (1926, pp. 114 ff.), see also Hammersley (1950) and Lord (1954 a).

The following formula connects the frequency function $f(v)$ of the distance between two random points in a plane convex region and the d. f. $C(v)$ of the length of random chords in the region

$$f(v) = 6v \int_v^\infty [1 - C(w)]\,dw \Big/ \int_0^\infty w^3 dC(w) \qquad (2.5.21)$$

The formula is a consequence of the relation existing between the density of pairs of points and the density of straight lines (Santaló 1953, pp. 16 – 19). The expressions (22) and (23) below have been derived by aid of (21).

The fr. f. of the distance between two random points in an equilateral triangle with side s is

$$\frac{8v}{s^2\sqrt{3}}[t_1(v/s) + t_2(v/s)] \qquad (2.5.22)$$

where

$$t_1(w) = \begin{cases} \pi - 4w\sqrt{3} + w^2(\sqrt{3} + 2\pi/3) & 0 < w < 1 \\ 0 & \text{otherwise} \end{cases}$$

$$t_2(w) = \begin{cases} 3\sqrt{12w^2 - 9} - (4w^2 + 6)\arccos(\sqrt{3}/2w) & \sqrt{3}/2 < w < 1 \\ 0 & \text{otherwise} \end{cases}$$

The corresponding frequency function for a regular hexagon with side s is

$$\frac{4v}{9s\sqrt{3}}h(v/s) \qquad (2.5.23)$$

with

$$h(w) = \begin{cases} 0 & w < 0,\ w > 2 \\ 3\pi - 4w\sqrt{3} + w^2(\sqrt{3} - \pi/3) & 0 < w < 1 \\ \pi(5 + w^2) - 3\sqrt{12w^2 - 9} - (4w^2 + 6)\arcsin\dfrac{\sqrt{3}}{2w} & 1 < w < \sqrt{3} \\ (2w^2 + 24)\left(\arcsin\dfrac{\sqrt{3}}{w} - \dfrac{\pi}{3}\right) - \sqrt{3}(w^2 + 6) + 10\sqrt{3w^2 - 9} & \sqrt{3} < w < 2 \end{cases}$$

Some distributions connected with points chosen along plane curves are listed in § 6.8.

2.6. Stationary stochastic set functions

A wide sense *stationary stochastic set function* can be defined in the following way. Random variables $\{Z(S)\}$ are given for all finite Borel sets in R_n. The following properties are assumed

$$Z(S_i \cup S_j) = Z(S_i) + Z(S_j) - Z(S_i \cap S_j)$$
$$E[Z(S_i)] = m\mu(S_i)$$
$$\text{Cov.}\,[Z(T_h S_i), Z^*(T_h S_j)] = \text{Cov.}\,[Z(S_i), Z^* S_j)]$$

Here μ is the Lebesgue measure and T_h is the translation operator, i. e. $T_h S$ is the set of points $x+h$ with $x \varepsilon S$. The covariance function shall be non-negative definite. If furthermore the covariance is unchanged by rotation, the stochastic set function $Z(S)$ is said to be (wide sense) *isotropic*.

The integrals considered in 2.5 furnish a first example of stationary set functions. However, there are important classes of stationary set functions which are not integrals of this type, e. g. the *point processes*.

For the general theory of stationary point processes in R_1, the reader is referred to fundamental papers by Wold (1949) and Bartlett (1954).

The simplest point process in R_n, the *n-dimensional* Poisson process, is constructed as follows. A random variable $Z(S)$, the number of "events" occurring in S, is attached to every set S. $Z(S)$ is assumed to have a Poisson distribution with mean $\lambda\,\mu(S)$, where the *intensity* λ is a positive number. Further, $Z(S_i)$ and $Z(S_j)$ shall be independent if S_i and S_j are disjunct. For arbitrary S_i and S_j it follows

$$\text{Cov. } [Z(S_i), Z(S_j)] = \lambda\,\mu(S_i \cap S_j) \tag{2.6.1}$$

Since the process is real the asterisk denoting the complex conjugate has been dropped here.

The Poisson process is an example of a stationary set function which is *orthogonal*. Processes of this kind are characterized by the property that $Z(S_i)$ and $Z(S_j)$ are uncorrelated if $S_i \cap S_j = 0$. The Poisson process is also isotropic.

An example of a more general kind is obtained by substituting an integrable stationary and positive stochastic process $\lambda(x)$ for the intensity. The ensuing set function $Z(S)$ will in the sequel be called a *stationary compound Poisson process*. The integral of λ over S is denoted $\Lambda(S)$. If a realization of $\lambda(x)$ is given, the conditional distribution of $Z(S_i)$ is a multidimensional Poisson distribution with

$$E[Z(S)] = \Lambda(S)$$

$$\text{Cov. } [Z(S_i), Z(S_j)] = \Lambda(S_i \cap S_j)$$

This way of constructing point processes is mentioned in Quenouille (1949), see also Bartlett (1954), and Thompson (1955).

Let the intensity function λ have mean m and cov. f. $\sigma^2 r(u)$. Then

$$E\ [Z(S)] = m\,\mu(S)$$

$$\text{Cov. } [Z(S_i), Z(S_j)] = m\,\mu(S_i \cap S_j) + \sigma^2 \int_{S_i}\int_{S_j} r(x-y)\,dx\,dy \tag{2.6.2}$$

The covariance thus is a sum of two terms, one of the type (1), and the other of the type belonging to integrals of stationary processes.

A further discussion of point processes is found in § 3.6 where some special cases are treated.

A generalization of the above set function is obtained by attaching to an "event" occurring in the point x a real quantity $z(x)$ and defining $Z(S)$ as

$$\Sigma z(x_i) \tag{2.6.3}$$

with summation over all events x_i in S.

Denote the mean value of

$$|z(x)|^2 \lambda(x)$$

by m, and let $\sigma^2 r(u)$ be the covariance function of the product $\lambda(x)z(x)$. It is then seen that the moment formula (2) is valid also for the stochastic set function (3). If $\lambda(x)$ is constant, (3) may be called a *generalized Poisson process*, cf. Feller (1943).

Still other types of stationary set functions may be of interest in the applied field. We may want to study e. g. a random variable $Z(S)$ defined as the total length of the intersection between S and a network of curves, generated by some random procedure. More generally $Z(S)$ may be the $(n-p)$-dimensional measure of the intersection between S and some enumerable set of $(n-p)$-dimensional subsets of R_n. Models of this type can be useful when treating estimation problems concerning lengths of random curves in R_2 or R_3, or areas of random surfaces in R_3. (For problems of estimating the length of networks in R_2, see Steinhaus 1954 and Matérn 1959.) However, we shall not discuss such stochastic set functions here. Yet, it may be remarked that the concept of stochastic process can be generalized so that a unified treatment including various types of random functions is possible; see papers by Ito (1954) and Yaglom (1957), in which the Schwartz theory of distributions is applied.

Chapter 3. Some particular models

3.1. Preliminaries

To amplify the purely formal treatment in Ch. 2 we shall now deal with some mechanisms that furnish sample functions of stationary processes. We shall not be concerned with general representations of the kind valid for all processes with a given cor. f. For such representations the reader is referred to Karhunen (1947). Instead some rather specific models will be investigated here.

Now a model must be of a simple and tangible type if it is meant to convey a clear idea of what a realization looks like and also to give possibilities of computing a sample function with sufficient accuracy.

Besides illustrating the theoretical concepts models of this kind can also give indications on the appropriate assumptions on correlation functions etc. in applications. Needless to say, a model must often be almost grotesquely oversimplified in comparison with the actual phenomenon studied.

Disregarding for the moment any connection with real phenomena, we can obtain sample functions corresponding to a given cor. f. in the following way (cf. Bartlett 1955, p. 163). Let r be a continuous cor. f. in R_n and denote the corresponding spectral d. f. by $F(x)$. Let further X be an n-dimensional random variable with d. f. $F(x)$, and let the one-dimensional random variable X_1 be independent of X and uniformly distributed over an interval of the length 2π. Then

$$z(x) = \exp(ix'X + iX_1) \tag{3.1.1}$$

has r as cor. f.

The model (1) is suggested by the general spectral representation of a cor. f. in C_n''. Turning to the isotropic case, the representation (2.3.3) suggests the model

$$z(x) = \exp(ix'XY + iX_1) \tag{3.1.2}$$

where X_1 has the same meaning as in (1), X is equidistributed on the surface of the unit n-sphere, and Y is an one-dimensional random variable with the ch. f. $g(v)$ of formula (2.3.6). X, Y, and X_1 are independent.

From the application point of view the above models must be classified as patologic. They may be useful e. g. if we want to compute complicated variances and covariances (cf. § 2.5) by experimental sampling ("Monte Carlo methods").

Mechanisms which are computationally simple and also bear some resemblance to reality may be obtained by aid of *the n-dimensional Poisson process* (see 2.6).

The Poisson process shall then furnish randomly located "centers" spreading their influence over a more or less restricted neighbourhood. Different types of mechanisms produce a large variety of stochastic processes. Many authors have used this approach; a short systematic account of such models in R_1 is found in Blanc-Lapierre & Fortet (1953, pp. 143 ff.). As remarked by these authors the dispersal from the center may be a disturbing effect produced by the mechanism used in observing the process.

3.2. Moving average model with constant weight function

Consider a Poisson process in R_n (cf. § 2.6) with intensity λ. The process produces randomly located centers. Denote by $dN(x)$ the number of centers in the volume element dx of R_n. Let further $q(x)$ be a quadratically integrable fix function. A stationary process $z(x)$ is then defined as

$$z(x) = \int_{R_n} q(x-y)\, dN(y) \tag{3.2.1}$$

We shall call q the *weight function* (cf. Zubrzycki 1957, p. 113) of the *moving average model* (1). The cov. f. of (1) is $\lambda c(u)$, where

$$c(u) = \int_{R_n} q(u+y)\, q^*(y)\, dy \tag{3.2.2}$$

The corresponding spectral d. f. is absolutely continuous with density

$$\text{const. } |\varphi(u)|^2 \tag{3.2.3}$$

where $\varphi(u)$ is the Fourier transform of q. Formula (3) is a consequence of Parseval's theorem.

It may be added that the quadratic integrability of q is not strictly sufficient for (1) to be meaningful. The mean value of $z(x)$ is formally deduced to be

$$\lambda \int_{R_n} q(x)\, dx$$

This integral may be divergent. However q may still be used as weight function if (1) is replaced by

$$\alpha + \int_{R_n} q(x-y)\, [dN(y) - \lambda dy]$$

As a first example, consider

$$q(x) = \begin{Bmatrix} B & \text{if } |x| < A \\ 0 & \text{otherwise} \end{Bmatrix} \tag{3.2.4}$$

Thus $z(x)/B$ is the number of centers within distance A from x. From (2) we obtain the isotropic cov. f.

$$\lambda B\, B^*\, V_n(A, A; v) \tag{3.2.5}$$

We have here used the notation $V_n(a, b; c)$ for the volume of the intersection of the two spheres

$$|x - x_1| < a \qquad\qquad |x - x_2| < b$$

where the distance $|x_1 - x_2|$ between the centers of the spheres equals c. V is the sum of two spherical segments. Thus

$$V_n(a, b; c) = C_n a^n \int_{\alpha}^{1} h_{n+2}(w)\, dw + C_n b^n \int_{\beta}^{1} h_{n+2}(w)\, dw \tag{3.2.6}$$

C_n denotes the volume of the unit n-sphere:

$$C_n = \pi^{n/2}/\Gamma(1 + n/2)$$

Furthermore, h is the function (2.3.4), and

$$\alpha = \frac{a^2 + c^2 - b^2}{2ac} \qquad \beta = \frac{b^2 + c^2 - a^2}{2bc}$$

It must be remembered that $h(w) = 0$ if $|w| > 1$. For some low values of n we find the following cor. f.'s corresponding to (4)

$n = 1 \quad r(2Av) = 1 - v$
$n = 2 \quad r(2Av) = 1 - (2/\pi)\left[v\sqrt{1 - v^2} + \arcsin v\right]$
$n = 3 \quad r(2Av) = 1 - 3v/2 + v^3/2$
$n = 5 \quad r(2Av) = 1 - 15v/8 + 5v^3/4 - 3v^5/8$

These expressions are valid for $|v| < 1$, otherwise $r(2Av) = 0$. The case $n = 2$ is found in Zubrzycki (1957, p. 115).

It is seen from (2.4.13) that the corresponding spectral density is (cf. 2.3.5)

$$[\Lambda_{n/2}(Av)]^2 \qquad (3.2.7)$$

Four additional weight functions and the corresponding correlation functions are listed in table 1, where also references are given to relevant formulae in Ch. 2. Since the functions are all isotropic, the distance v is used as argument, as in Ch. 2.

Weight functions corresponding to a given cor. f. can be found by aid of the Fourier transforms and (3).

Table 1. Weight functions and corresponding correlation functions.
$[k = (n-2)/2 \quad s > 0 \quad a > 0]$

Weight function	Correlation function	Formulae in § 2.4
$\exp(-2a^2v^2)$	$\exp(-a^2v^2)$	2
$(1 + 4v^2/a^2)^{-\frac{n+1}{2}}$	$(1 + v^2/a^2)^{-\frac{n+1}{2}}$	5
$\Lambda_{k+\frac{s+1}{2}}(av)$	$\Lambda_{k+s}(av)$	12
$v^t K_t(av)$	const. $v^s K_s(av)$ with $s = 2t + n/2$	7

We note some special cases of the last example of table 1. If $s = {}^1/_2$, it is seen that $r(v) = \exp(-av)$ corresponds to the weight function

$$v^{-\frac{n-1}{4}} K_{\frac{n-1}{4}}(av) \tag{3.2.8}$$

For $n = 1$ and $n = 2$ the weight functions are

$$K_0(av) \quad \text{and} \quad v^{-1/4} K_{1/4}(av)$$

respectively, i. e. weight functions that are infinite at the origin. (A table of $K_{1/4}$ is found in Carsten & McKerrow 1944). If the weight function $K_0(av)$ is chosen for all n, we find the cor. f.'s

$$2\,(av/2)^{n/2} K_{n/2}(av) \,/\, \Gamma(n/2) \tag{3.2.9}$$

This is of some interest in conjunction with Whittle's discussion of the two-dimensional analogue of a Markoff process, see the remark attached to (2.4.9). If the weight function is $\exp(-av)$ for all n, the corresponding cor. f.'s are given by (9) with n replaced by $n+2$.

From the isotropic models we can derive models with "elliptic" cor. f.'s by replacing the isotropic weight function $q(v)$ by $q(\sqrt{x'Ax})$ where A is a positive definite matrix. The cor. f. $r(v)$ corresponding to $q(v)$ is then changed into $r(\sqrt{x'Ax})$, cf. (2.4.16). In this context we note the following generalization of the scheme (4):

$$q(x) = \begin{Bmatrix} B & x \varepsilon S \\ 0 & \text{otherwise} \end{Bmatrix} \tag{3.2.10}$$

Here S is a set of finite measure in R_n. The spectral intensity of the corresponding process is proportional to the ch. f. of $X - Y$, where X and Y are independently chosen with uniform distribution over S.

It may be remarked that (1) can be described as a linear filter with q as a "transient response function", see Grenander & Rosenblatt (1956, p. 49). We note also that a model of this type can be constructed to every cor. f. that has an absolutely continuous spectrum, see Karhunen (1947).

We finally note that in the case of stochastic processes in the proper sense ($n = 1$), the weight function should usually not be chosen as isotropic; it would mostly be realistic to have $q(x) = 0$ for $x < 0$, i.e. a one-sided filter.

3.3. Moving average model with stochastic weight function

The model considered in the previous section can be written

$$z(x) = \Sigma\, q(x - y_j) \tag{3.3.1}$$

where $y_1, y_2, \ldots$, are the centers produced by the Poisson process, arranged e. g. in ascending order of distance from the origin of R_n. Suppose now that to each center is selected at random a weight function out of a certain family of functions $\{q(x;t)\}$, where each function is quadratically integrable over R_n. Let the random variable t vary in the space T according to the probability function $P(t)$. Thus, to each center y_j a sample value t_j of this random variable is chosen. The t-values attached to the different centers are assumed to be independent.

For fixed x, $q(x;t)$ is a random variable. We suppose that its first two moments are finite. The following notation is used

$$q_0(x) = E[q(x;t)]$$

$$p(u;x) = E[q(x+u;t)\, q^*(x;t)]$$

It is assumed that $q_0(x)$ and $p(0;x)$ are integrable over R_n.

The process with stochastic weight function can now be written in a form analogous to (1)

$$z(x) = \sum_j q(x - y_j; t_j) \tag{3.3.2}$$

We can also interpret $z(x)$ as the convolution of processes with fixed weight functions; the process with weight function $q(x;t)$ having the intensity $\lambda dP(t)$. Thus the following alternative to (2) is obtained

$$z(x) = \int_{R_n \times T} q(x - y; t)\, dN(y, t) \tag{3.3.3}$$

where $dN(y, t)$ is the number of centers produced in the interval $dy\, dt$. The expected number is given by

$$E[dN(y, t)] = \lambda\, dy\, dP(t)$$

Clearly $N(y, t)$ forms "a process with independent increments". We now introduce

$$M(t) = \lambda P(t) \tag{3.3.4}$$

and call $M(t)$ the *mixing function.*

The approach (2) may be generalized. We may admit unbounded mixing functions. In this case some further restrictions must be placed on the family of weight functions so that $q_0(x)$ and $p(0;x)$ become integrable over R_n. Extending the notation of (3.2.2) we now introduce

$$c(u; t) = \int_{R_n} q(u - y; t)\, q^*(y, t)\, dy \tag{3.3.5}$$

The cor. f. of $z(x)$ is

$$r(u) = \text{const.} \int_T c(u; t)\, dM(t) \tag{3.3.6}$$

The corresponding spectral density is (cf. 3.2.3)

$$f(u) = \text{const.} \int_T |\varphi(u; t)|^2 dM(t) \qquad (3.3.7)$$

where $\varphi(u;t)$ is the Fourier transform of $q(u;t)$.

As first example we consider the family

$$q(v; t) = t^{-n} \exp(-v^2/2t^2) \qquad (3.3.8)$$

where t is restricted to real numbers. As mixing function we choose $M(t)$, defined by

$$\frac{dM(t)}{dt} = \begin{cases} t^{n+2\nu-1} \exp(-t^2b^2) & t > 0 \\ 0 & t < 0 \end{cases} \qquad (3.3.9)$$

Here ν must be > 0 to make $p(0;x)$ integrable. With $|u| = w$

$$\varphi(u;t) = \text{const.} \exp(-t^2w^2/2)$$

Applying (7) we obtain

$$f(u) = \text{const.} (1 + w^2/b^2)^{-(\nu+n/2)}$$

Thus (cf. 2.4.5 – 2.4.7)

$$r(v) = 2(bv/2)^\nu K_\nu(bv)/\Gamma(\nu) \qquad (3.3.10)$$

Let us still retain the weight functions (8) but replace (9) by

$$\frac{dM(t)}{dt} = \begin{cases} t^{n-2\nu-1} \exp(-a^2/4t^2) & t > 0 \\ 0 & t < 0 \end{cases} \qquad (3.3.11)$$

with $\nu > 0$. For $\nu \leqslant n/2$ the mixing function is unbounded. We now apply (6) to the functions (11) and

$$c(u; t) = t^{-n} \exp(-v^2/4t^2) \,.\, \pi^{n/2}$$

Then

$$r(v) = (1 + v^2/a^2)^{-\nu} \qquad (3.3.12)$$

The two forms (10) and (12) can be comprised into one expression if we choose (for positive t)

$$\frac{dM(t)}{dt} = t^{n+2\nu-1} \exp(-t^2b^2 - a^2/4t^2)$$

Then (cf. the derivation of 2.4.6)

$$r(v) = 2(1 + v^2/a^2)^{(\nu-s)/2} (\theta/2)^s K_s(\theta)/\Gamma(s) \qquad (3.3.13)$$

with $\theta = b\sqrt{a^2+v^2}$, $s = |\nu|$. The case (10) corresponds to the choice $a=0, \nu>0$. To get (12) we must take $\nu<0$ and make $b\to 0$.

A different type of models is obtained by choosing (cf. 3.2.11)

$$q(x;t) = \begin{cases} B_t & x \in S_t \\ 0 & \text{otherwise} \end{cases} \tag{3.3.14}$$

where the complex number B_t and the set S_t in R_n are attached to the random variable t. The corresponding cor. f. has the spectral density

$$f(u) = \text{const.} \int_T |B_t|^2 \mu^2(S_t) \,|\, \varphi_t(u) \,|^2 \, dM(t) \tag{3.3.15}$$

In (15) $\mu(S_t)$ denotes the volume of S_t, $\varphi_t(u)$ is the ch. f. of the rectangular coordinates of a point chosen at random in S_t. The formula can be condensed into

$$f(u) = \text{const.} \int_T |\, \varphi_t(u) \,|^2 \, dL(t) \tag{3.3.16}$$

where $L(t)$ is non-decreasing. If a tractable model corresponding to a particular cor. f. is wanted, $dL(t)$ may be partitioned into factors λ, $|B_t|^2$, $\mu^2(S_t)$ and $dP(t)$ in any manner that gives computational ease.

A paper by Hammersley and Nelder (1955) is devoted to the special case where the sets S_t of (14) are spheres. If t is the diameter of the sphere S_t in R_n, it can be seen from (16) and (3.2.7) that the spectral density is

$$\text{const.} \int_0^\infty [\Lambda_{n/2}(tw/2)]^2 \, dL(t) \tag{3.3.17}$$

The cor. f. is (cf. 3.2.6)

$$r(v) = \text{const.} \int_v^\infty t^n \, dL(t) \int_{v/t}^1 h_{n+2}(w) \, dw \tag{3.3.18}$$

Differentiating, we get

$$-r'(v) = \text{const.} \int_v^\infty t^{n-1} \, h_{n+2}(v/t) \, dL(t) \tag{3.3.19}$$

To obtain an example we start from the following integral in Bessel functions (a transformation of formula 6, p. 417 in Watson 1944)

$$\int_v^\infty t^{n-1} h_{n+2}(v/t) \, t^{1-s} K_s(at) \, dt = \text{const.} \, v^p K_p(av)$$

where

$$p = (n+1-2s)/2$$

From this formula one cor.f. of the type (2.4.7) can be obtained, viz. by putting $s = n/2$. It can be seen that the cor.f. $\exp(-av)$ corresponds to the choice

$$\frac{dL(t)}{dt} = K_{n/2}(at)\, t^{1-n/2}$$

Hammersley & Nelder have treated the special cases $n = 1$, 2, and 3.

It follows from (18) that $r(v)$ is a linear functional of L. We therefore conclude that any completely monotonic cor. f. (2.4.10) can be obtained by the Hammersley-Nelder procedure.

As a consequence of (19)

$$r'(0) < 0 \tag{3.3.20}$$

It is seen that most of the cor.f.'s given as examples in 2.4 do not have the property (20). We note for example that $\exp(-av)$ is the only cor.f. of the type (2.4.7) with a finite negative $r'(0)$. [If the index ν in (2.4.7) is $> 1/2$, $r'(0)$ is 0; for $\nu < 1/2$, $r'(0)$ attains the negative infinity.]

Expressed in the present notation, the chief problem solved by Hammersley & Nelder is to find the function L corresponding to a given cor.f. in R_n. Now the relation between r' and L is of the same nature as the connection between a cor.f. and the ch.f. of the corresponding radial distribution, cf. § 2.3, esp. (2.3.6) and (2.3.10 – 12). Disregarding for the moment the statistic properties of L and r, we write down the formal solution of the integral equation (18). Denoting more fully the right hand side of (18) as $r_n(v)$, we get by differentiating (19)

$$-r''_n(v) = \text{const.}\ v\, r'_{n-2}(v) \tag{3.3.21}$$

If $r_n(v)$ is given, the corresponding r_0 or r_1 can be found by means of (21). The function L is then obtained by one or the other of the formulas

$$\frac{dL(t)}{dt} = \frac{\text{const.}}{t} \int_t^\infty \frac{v\, d[v r'_0(v)]}{\sqrt{v^2 - t^2}} \tag{3.3.22}$$

$$\frac{dL(t)}{dt} = \text{const.}\ r_1''(t) \tag{3.3.23}$$

Using (21) and (22) we obtain for example

$$\frac{dL(t)}{dt} = \frac{\text{const.}}{t} \int_t^\infty \frac{v\, dr''_2(v)}{\sqrt{v^2 - t^2}} \tag{3.3.24}$$

It should be added that Hammersley & Nelder use a special device to assure that an exact realization of a process over a bounded region can be obtained in a finite number of steps.

One way to obtain models in R_n with stochastic weight functions is to form "*subprocesses*" of processes in R_{n+m} with fixed weight function. Let $x=(x_1, x_2)$ be a point in R_{n+m}, x_1 signifying the n first coordinates, and x_2 the remaining m coordinates. Assume that $z(x)$ is a moving average process in R_{n+m} with the fixed weight function $q(x)$. A subprocess in R_n can then be defined as

$$z_1(x_1) = z(x_1, 0) = \int_{R_{n+m}} q(x_1 - y_1; -y_2)\, dN(y_1, y_2) \qquad (3.3.25)$$

The process is here written in the form (3). An example of the Hammersley-Nelder type is obtained if

$$q(x) = \begin{cases} 1 & |x| < A/2 \\ 0 & \text{otherwise} \end{cases}$$

We can then express z_1 in the form (2)

$$z_1(x_1) = \sum_j q(x_1 - y_j; t_j)$$

with

$$q(x_1; t) = \begin{cases} 1 & |x_1| < t/2 \\ 0 & \text{otherwise} \end{cases}$$

where t is a real variable ranging from 0 to A. The mixing function is

$$M(t) = \text{const.}\ [1 - (1 - t^2/A^2)^{m/2}]$$

The derivative of the cor. f. is found by means of (19)

$$r'(v) = \text{const.}\ (1 - v^2/A^2)^{(m+n-1)/2}$$

Hence the cor.f. has a spectral distribution of the type (3.2.7) which also is evident from the way the process was constructed.

Some short indications about possible applications of the models considered will now be given.

It will first be noted that many observations in practice are restricted to subspaces. E.g. a three-dimensional manifold is observed by means of two-dimensional sections, see the treatment of the "corpuscle problem" in Wicksell (1925, 26) and Thompson et al. (1932). Further, two-dimensional regions are often observed by means of one-dimensional line transects.

A special interpretation of the model (14) should also be mentioned. The set S_t attached to a center can be thought of as the *visibility region* of the center. If further $B_t = 1$, $z(x)$ is the number of centers visible from x. We may in this connection think of the Bitterlich "Winkelzahl" method of estimating the basal area of a forest stand, for references see Bitterlich (1956).

If $m=1$ and $q=0$ for positive values of y_2 in (25), we may think of y_2 as signifying *time*. Centers appear in R_n with constant intensity in every time-interval. At the time of "birth" the influence is spreading according to the function

$$q(x_1 - y_1; 0)$$

After an elapse of y_2 time units the pattern is changed to

$$q(x_1 - y_1; -y_2)$$

If a real explanation of any phenomenon showing spatial varation is intended, time must be introduced in the model, as mentioned e.g. by Whittle (1956).

Finally, we note that if $n = 1$ and the influence from the centers is spreading out in one direction only, we get a number of well-known stochastic processes in the proper sense. If

$$q(x; t) = \begin{cases} 1 & 0 < x < t \\ 0 & \text{otherwise} \end{cases}$$

and t is exponentially distributed, a simple process is obtained which is of interest e.g. in studies of telephone traffic, see Feller (1950, p. 377).

3.4. Distance models

Several types of processes can be obtained from

$$z(x) = \sum_i q(x - y_i)$$

if the q's are allowed to be *mutually dependent* random functions. We shall here mention briefly some rather special cases that will be called "distance-models" since the distances from x to the centers $(y_1, y_2, \ldots)$ enter in the definition of $z(x)$.

Let the ordered sequence of distances from the point x to the centers produced by a Poisson process in R_n be

$$\alpha_1(x) \leqslant \alpha_2(x) \leqslant \ldots$$

A model of the distance type is

$$z(x) = q[\alpha_1(x), \alpha_2(x), \ldots]$$

where q is a fixed function.

A simple example is

$$q(x) = \begin{Bmatrix} 1 & \textit{if } \alpha_1(x) \leqslant A \\ 0 & \text{otherwise} \end{Bmatrix} \tag{3.4.1}$$

Thus $z(x) = 1$ if at least one center is within distance A from x.

In the case $n=2$ the model (1) is closely related to a "bombing problem" studied by several writers, see references in Solomon (1953). Bombs are dropped at random over a region. Each bomb devastates the area within distance A from the point of impact. Of interest are the properties of the total area affected.

By geometrical considerations

$$E[z(x)] = 1 - \exp[-\lambda C_n A^n] \qquad (3.4.2)$$

where as before λ is the intensity of the Poisson process, and C_n is the volume of the unit n-sphere (see 3.2). Further, the covariance function is

$$\text{Cov.}[z(x_1), z(x_2)] = \exp(-2\lambda C_n A^n)\{\exp[\lambda V_n(A, A; v)] - 1\} \qquad (3.4.3)$$

where $v = |x_1 - x_2|$ and (cf. 3.2.6)

$$V_n(A, A; v) = 2 C_n A^n \int_{v/2A}^{1} h_{n+2}(w)\, dw \qquad (3.4.4)$$

is the volume common to two spheres with the radius A and the centers v apart. The process partitions R_n into two random sets, one with $q=1$, the other with $q=0$. We shall therefore return to this example when discussing the random sets models in the next section.

In conjunction with certain sampling survey problems it would be of interest to consider models of the type

$$z(x) = [\alpha_\nu(x)]^\mu \qquad (3.4.5)$$

for some low values of ν and μ (e.g. $\nu \leqslant 4$, $\mu = -2, 1, 2$). Methods have been developed for using observations of this kind to estimate the number of individuals (plants, larvae, etc.) per unit area, see Bauersachs (1942), Cottam & Curtis (1956), Matérn (1959). The methods proposed are based on the assumption that the location of the individuals follows a Poisson process. Similarly such observations can be useful when testing the Poisson scheme against some "contagious" or other model, cf. Skellam (1952), Strand (1954).

In studying the statistical properties of (5), one seems to run into rather laborious and unwieldy computations. It is only the mean value that can be immediately obtained

$$E\{[\alpha_\nu(x)]^\mu\} = \Gamma(\nu + \mu/n)/(\lambda C_n)^{\mu/n}\Gamma(\nu) \qquad (3.4.6)$$

For the simplest case, $\nu = \mu = 1$, we further find by geometrical arguments

$$P[\alpha_1(x_1) > a;\ \alpha_1(x_2) > b] = \exp[-\lambda U(a, b; v)] \qquad (3.4.7)$$

where

$$U(a, b; v) = C_n a^n + C_n b^n - V_n(a, b; v)$$

is the volume of the union of the two spheres

$$|x - x_1| < a \qquad\qquad |x - x_2| < b$$

with $|x_1 - x_2| = v$. V_n is given by (3.2.6). Using (7) it can be seen that

$$E[\alpha_1(x_1)\,\alpha_1(x_2)] = \int_0^\infty\int_0^\infty \exp[-\lambda\, U(a, b; v)]\, da\, db \tag{3.4.8}$$

An expression for the cov.f. of $\alpha_1(x)$ can be obtained from (6) and (8). In the case $n = 1$ $\alpha_1(x)$ is seen to have the cor.f.

$$(1 + 2\lambda v)\exp(-2\lambda v)$$

The same cor.f. was also obtained by the procedure of 3.2 with the fixed weight function $\exp(-2\lambda v)$.

In some surveys (e.g. of market conditions for wood) it may be of more interest to obtain data on "economic" distances (along roads, waterways, etc.) than to find the mathematical distances. The correlation structure of the corresponding process must however be connected with the structure of the process $\alpha_1(x)$ considered here. In other cases information is needed about distances from a given point to the nearest point on a network of roads, or waterways, etc., see v. Segebaden (1964). To treat the corresponding stochastic process, we need first a basic mechanism generating a family of random curves in R_2; cf. the end of section 2.6.

3.5. Models of random sets

This section treats processes which can take on only the two values 0 and 1. We interpret the set of points x with $z(x) = 1$ as a "random set", S, in R_n. Thus $z(x)$ is the characteristic function of S in the set-theoretical meaning.

Throughout the section we assume

$$\begin{aligned} P[z(x) = 1] &= E[z(x)] = p \\ P[z(x) = 0] &= q = 1 - p \end{aligned} \tag{3.5.1}$$

The cov.f. is denoted $c(x)$ or (in the isotropic case) $c(v)$, where $v = |x|$, as before. The cor.f. is

$$r(x) = c(x)/pq$$

In the previous section one particular model of this type was mentioned, i.e. the model (3.4.1) corresponding to a bombing problem.

The following model is also based on the Poisson process. To each center $y \in R_n$ produced by a Poisson process in R_n is attached a "cell" Y consisting of those points which have y as nearest center. It is then decided by a random procedure with parameters given by (1) whether all points of Y shall have

$z(x) = 1$ or $z(x) = 0$. As above the points with $z(x) = 1$ form the "random set" S. To complete the definition, S may be assumed to be closed. Further the random procedures carried out for different Y's will be supposed to be independent.

Both the above processes are isotropic. In the latter example the cor.f. $r(v)$ equals the probability that two points at distance v apart shall belong to the same set. This probability is

$$r(v) = \lambda \int_{R_n} \exp[-\lambda U(|y - x_1|, |y - x_2|; v)]\, dy \tag{3.5.2}$$

where U is the same function as in (3.4.7), x_1 and x_2 are two points at the mutual distance v. For $n = 1$ we have

$$r(v) = \exp(-2\lambda v)\,(1 + \lambda v) \tag{3.5.3}$$

For $n > 1$, (2) can be written as a double integral. Assume that x_1 and x_2 are the two points

$$x_1 = (-v/2, 0, 0, \ldots, 0)$$
$$x_2 = (v/2, 0, 0, \ldots, 0)$$

Write y in (2) as $y = (y_1, y_2, \ldots, y_n)$ and introduce two new variables by

$$t = y_1 \qquad w = (y_2^2 + y_3^2 + \ldots + y_n^2)^{1/2}$$

Then

$$r(v) = 2\lambda(n-1)\, C_{n-1} \int_0^\infty w^{n-2} dw \int_0^\infty \exp[-\lambda U(a, b; v)]\, dt \tag{3.5.4}$$

where

$$a^2 = (t + v/2)^2 + w^2 \qquad b^2 = (t - v/2)^2 + w^2$$

and C_{n-1} is the volume of the unit sphere in R_{n-1}.

It will now be briefly indicated that the derivative

$$\left[\frac{dc(v)}{dv}\right]_{(v=0)} = c'(0)$$

has a rather tangible meaning for the isotropic random set processes.

Let the boundary of the random set S be an $(n-1)$-dimensional area of finite surface-content within every bounded part of R_n. Denote by $S(\varepsilon)$ the set of points within distance ε from the boundary of S, and by $P(\varepsilon)$ the probability that an arbitrary point belongs to $S(\varepsilon)$. If $x_0 \in S(\varepsilon)$, a certain part of the sphere $|x - x_0| < \varepsilon$ consists of points x such that $z(x) \neq z(x_0)$. For a small ε this subset is approximately a spherical segment. The perpendicular distance from x_0 to the segment can be regarded as rectangularly distributed between 0 and ε. The expected n-dimensional volume of the segment, $A(\varepsilon)$ (say), is then approximately given by

$$A(\varepsilon) \approx (1/\varepsilon) \int_0^{\varepsilon} C_n \varepsilon^n dt \int_{t/\varepsilon}^{1} h_{n+2}(w)\, dw = C_n \varepsilon^n / (n+2)\, B\left(\frac{n+3}{2}, \frac{1}{2}\right)$$

We then express the product $P(\varepsilon) A(\varepsilon)$ in the covariance function. We first note that

$$P[z(x) \neq z(x_0)] = 2[c(0) - c(v)] \approx -2\, v\, c'(0)$$

with $v = |x - x_0|$. Hence

$$P(\varepsilon) A(\varepsilon) = \int P[z(x) \neq z(x_0)]\, dx \approx -2\, c'(0) \int v dx$$

where the integration is carried out over the sphere

$$v = |x - x_0| < \varepsilon$$

Transforming the integral we obtain

$$P(\varepsilon) A(\varepsilon) \approx -2 c'(0) \int_0^{\varepsilon} v C_n dv^n = -2 c'(0)\, C_n \frac{n \varepsilon^{n+1}}{n+1}$$

Now for small ε

$$P(\varepsilon) \approx 2\, \varepsilon\, T$$

where T is the expected surface-content of S within a unit volume of R_n.

Eliminating $P(\varepsilon)$ and $A(\varepsilon)$ from the above relations, and passing to the limit, we finally get

$$T = -2 c'(0) \sqrt{\pi}\, \Gamma\left(\frac{n+1}{2}\right) \Big/ \Gamma\left(\frac{n}{2}\right) \qquad (3.5.5)$$

Of course, these geometrical considerations do not constitute a rigorous proof of (5). It would be of interest to describe more exactly the class of surfaces, which have the properties implied in the above expressions. This is, however, beyond the scope of the present investigation. Yet, it may be mentioned that the definition of surface area as the limit of the quotient $V(\varepsilon)/2\varepsilon$ [where $V(\varepsilon)$ is the volume of $S(\varepsilon)$] seems first to have been given by Minkowski in 1901, see Cesari (1956, p. 77).

For low values of n (5) gives:

$$n = 1 \qquad T = -2\, c'(0)$$
$$n = 2 \qquad T = -\pi\, c'(0)$$
$$n = 3 \qquad T = -4\, c'(0)$$

Here, T signifies the number of boundary points ($n=1$), the length of boundary lines ($n=2$), and the area of the bounding surface ($n=3$). The relation for $n=1$ is implicitly contained in a formula given by Williams (1956, p. 143, formula 9).

It is seen that (5) is valid also for the process (3.2.1) if the weight function is (3.2.10) with $B=1$. Then $z(x)$ is constant within certain regions and changes its value by the amount 1 when passing a boundary. An application to level surfaces of a process with continuous realizations is indicated in § 4.3.

The formula may also be generalized in other ways. Suppose e.g. that R_n is divided into cells as in the second introductory example of this section. Let all points x in a cell receive the same "score" $z(x)$ chosen from among the values $(a_1, a_2, \ldots)$ with probabilities $(p_1, p_2, \ldots)$. Let further the score given to the points in a cell be independent of the scores of all other cells. Denoting by T_1 the expected surface content (per unit volume)of the boundaries between cells, we find

$$T_1 = -r'(0)\sqrt{\pi}\,\Gamma\left(\frac{n+1}{2}\right)\Big/\Gamma\left(\frac{n}{2}\right) \tag{3.5.6}$$

We can also prove the inequality

$$c(v) > c(0) + v\,c'(0)$$

which shows that *the cov.f. of an isotropic random set model is (downward) convex to the right of the origin.* Proof. Let θ denote the expected number of boundary points on a line of unit length. Now

$$z(x+u) \neq z(x)$$

if there is an odd number of boundary points between x and $x+u$. The probability of this event is less than the expected number of boundary points on the line, which is θv, where $v=|u|$. Hence

$$2\,[c(0)-c(v)] < \theta v$$

Dividing by v, and letting $v \to 0$, we have

$$\theta = -2\,c'(0) \tag{3.5.7}$$

and the inequality follows.

Combining this result with (5), we obtain

$$T = \theta\sqrt{\pi}\,\Gamma\left(\frac{n+1}{2}\right)\Big/\Gamma\left(\frac{n}{2}\right) \tag{3.5.8}$$

This formula connects the average number of intersections between the boundary of the random set and a fixed system of curves with the average surface-content of the boundary. Here, average means "per unit length" and "per unit volume", respectively. Formula (8) is valid also in the non-isotropic case, if θ denotes the average obtained when the system of curves is moved in a random manner in R_n; for a precise formulation, see books on integral geometry, e.g. Santaló (1953). For $n=2$

$$T=\theta\,\pi/2$$

an expression relevant in Buffon's needle problem.

We shall now consider some examples. We start with the model (3.4.1). Differentiating (3.4.3) with respect to v and applying (5), it is seen that

$$T=n\lambda C_n A^{n-1}\exp\left(-\lambda C_n A^n\right) \tag{3.5.9}$$

For $n=2$

$$T=2\,\pi\,\lambda\,A\,\exp(-\pi\,\lambda\,A^2) \tag{3.5.10}$$

The maximum value of T, involving "maximal patchiness", is obtained for fixed A if $\lambda=1/\pi A^2$. Then (3.4.2) gives

$$E[z(x)]=1-1/e=0.632$$

As indicated by this result and intuitively clear, the model (3.4.1) is unsymmetric in so far as the region with $z(x)=0$ (the complement of S) cannot be obtained by a procedure of the same nature as the one used in constructing S.

The second model considered in this section is symmetric in the above respect. We first note the value

$$c'(0)=-\lambda^{1/n}pq\frac{\left[\Gamma\left(\frac{n+2}{2}\right)\right]^{2-\frac{1}{n}}\Gamma\left(2-\frac{1}{n}\right)\Gamma(n)}{\Gamma\left(n-\frac{1}{2}\right)\left[\Gamma\left(\frac{n+1}{2}\right)\right]^2}\cdot\frac{n}{2}$$

The expression can be found from (4) after somewhat lengthy but trivial computations. Applying (5) we find for $n=1$ and $n=2$

$$n=1 \qquad T=2\,\lambda\,pq$$
$$n=2 \qquad T=4\sqrt{\lambda}\,pq$$

For $n=2$, (6) further gives $T_1=2\sqrt{\lambda}$. This result implies that the average perimeter per cell is $4/\sqrt{\lambda}$, whereas the average area per cell is $1/\lambda$. Thus the relation between area and perimeter in the average cell is the same as that in a square.

In the case $n=1$ the random set model can be interpreted as a type of a "renewal process", a system developing in time and alternating between the two states $z=0$ and $z=1$. In other words, an infinite sequence of "1-intervals" and "0-intervals" covers the time-axis (cf. Wold 1949, esp. p. 77). We shall now consider a model that is somewhat more general than the one with cor.f. (3).

Assume that the length of a "1-interval" has the d.f. $F_{1,0}(x)$, and that the length of a "0-interval" is distributed according to $F_{0,1}(x)$. Suppose further that after a start in the infinitely remote past the states alternate in such a way that the duration of a particular 0- or 1-interval is independent of the previous history of the system. (A slightly more general model is obtained if we assume that the length of a 0-interval is dependent on the length of the immediately preceding 1-interval, cf. Pyke 1958.)

Let $F_{m,n}(x)$ denote the d.f. of the sum of $m+n$ independent random variables, of which m have the d.f. $F_{1,0}$ and n have the d.f. $F_{0,1}$. Let $\mu(m, n)$ be the expectation of the corresponding random variable, which we suppose to be finite. Also, suppose $F_{1,0}(0)=F_{0,1}(0)=0$.

The probability that $z(t)$ equals 1 is (see Pyke 1958)

$$\mu(1, 0)/\mu(1, 1) \tag{3.5.11}$$

When $z(t)=1$, the conditional d.f. of the remaining duration of the 1-interval will be denoted $J(x)$. We have

$$dJ(x)/dx=[1-F_{1,0}(x)]/\mu(1, 0) \tag{3.5.12}$$

The conditional probability of $z(t+x)=1$, given that a 0-interval starts in t, is

$$P(x)=\sum_{s=1}^{\infty}[F_{s-1,s}(x)-F_{s,s}(x)] \tag{3.5.13}$$

The formula presupposes that $x>0$. Using (11) – (13) we then have if $v>0$

$$E[z(t)\,z(t+v)]=\frac{\mu(1, 0)}{\mu(1, 1)}[1-J(v)+\int_0^v P(v-u)\,dJ(u)] \tag{3.5.14}$$

Denoting the convolution of J and $F_{m,n}$ by $G_{m,n}$, we obtain from (13) and (14)

$$E[z(t)\,z(t+v)]=\frac{\mu(1, 0)}{\mu(1, 1)}\{1-J(v)+\sum_{s=1}^{\infty}[G_{s-1,s}(v)-G_{s,s}(v)]\} \tag{3.5.15}$$

By aid of (14) or (15) expressions for the cov.f. and cor.f. can be found.

An example, which is closely connected with the Poisson process, is obtained if we choose type III distributions for $F_{0,1}$ and $F_{1,0}$. We use the notation

$$k_m(x, \alpha)=\alpha^m x^{m-1}\exp(-\alpha x)/\Gamma(m)$$

$$K_m(x, \alpha)=\int_0^x k_m(u, \alpha)\,du$$

Assume now $F_{1,0}=K_m(x, \alpha)$ and $F_{0,1}=K_n(x, \beta)$, where m and n are positive integers. The corresponding system can be described in terms of two mutually independent Poisson-processes with the intensities α and β, respectively. A 1-interval starts at a certain event in the "β-process" and prevails until m events have happened in the "α-process". The subsequent 0-interval continues until n events have appeared in the "β-process".

We have $\mu(1, 0) = m/\alpha$ and $\mu(0, 1) = n/\beta$. Further

$$J(x) = (1/m) \sum_{r=1}^{m} K_r(x, \alpha)$$

Using $K_{r,s}(x, \alpha, \beta)$ for the convolution of $K_r(x, \alpha)$ and $K_s(x, \beta)$ we find from the well-known addition properties of type III variables

$$G_{r,s}(x) = (1/m) \sum_{j=1}^{m} K_{mr+j,\, ns}(x, \alpha, \beta) \tag{3.5.16}$$

In the simplest case, $m = n = 1$, (15) and (16) give

$$c(v) = \frac{\alpha\beta}{(\alpha+\beta)^2} \exp[-v(\alpha+\beta)] \tag{3.5.17}$$

If $\alpha = \beta$, it can be seen that for all integral m and n

$$c(v) = \frac{1}{(m+n)^2} \sum_{k=1}^{m+n-1} \exp(-2\alpha v \sin^2 \varphi_k) \cos[\alpha v \sin 2\varphi_k] \left[\frac{\sin m\varphi_k}{\sin \varphi_k}\right]^2 \tag{3.5.18}$$

where φ_k denotes $\pi k/(m+n)$. Thus $c(v)$ is a weighted sum of damped oscillations with varying wave-lengths. The easiest way to obtain (18) seems to be by expanding into partial fractions the Fourier-Stieltjes transform of (15).

Some model of this kind seems to give a reasonably realistic picture of the course of an activity where two types of work succeed one another. It will be used to illustrate some sampling problems connected with time studies, see § 6.2. It is evident that the models just considered comprise cases of rather high irregularity (low values of m and n) and also cases where the two activities alternate in a very regular manner (high values of m and n). At the extreme we have the case where the 1-interval always has duration a and the length of a 0-interval always is b. Assuming $a < b$, we find the cor.f.

$$r(v) = \begin{cases} 1 - (a+b)x/ab & 0 \leqslant x < a \\ -a/b & a \leqslant x < b \\ 1 - (a+b)(a+b-x)/ab & b \leqslant x < a+b \end{cases} \tag{3.5.19}$$

The cor.f. is further periodic with period $(a+b)$.

3.6. Models of randomly located points

Many authors have treated the distribution of points located in space according to one or the other random procedure. Research has mainly concentrated on the frequency distribution of volume units containing 0, 1, 2, . . . points (or rather areal units, since the studies have chiefly been devoted to distributions in the plane). A recent paper, containing references to earlier contributions, is Gurland (1958).

For the present investigation this approach is unsufficient, since it does not elucidate the covariance properties of the process. Also, it does not give satisfactory clues to the mechanism underlying the spatial arrangement, as pointed out by Skellam (1952, p. 347).

Papers devoted to more completely specified models are rather limited in number. The pioneering contribution by Neyman (1939), and some subsequent papers, as Thomas (1949), Skellam (1952), and Thompson (1954, 1955) should be mentioned. (For comments on Neyman's paper, see Feller 1943, and Skellam 1958.) Also the general exposition in Bartlett (1954), referred to earlier, should be mentioned in this context. The background of these papers is a discussion of distributions encountered in ecology. As seen from the reviews by Neyman & Scott (1958) and Fox (1958), already mentioned, similar models can be relevant also in astronomy.

Most of the models met with in the literature are based upon a primary Poisson distribution of "centers". Each center is surrounded by a cluster of "satellites". The number of satellites in the cluster and their location result from some random experiment. The experiment is carried out independently for each center, also the location of a particular satellite in respect to its center is independent of the corresponding location of the rest of the cluster.

Let $Z(S)$ denote the number of satellites in a region S of R_n. $Z(S)$ is a stationary stochastic set function in the meaning of § 2.6.

Denote by λ the intensity in the primary Poisson process. Assume that the number of satellites belonging to a particular cluster has finite expectation m and finite variance τ^2. Let $f(x-y)$ be the probability density of the rectangular coordinates (x) of a satellite belonging to a center located in y. We introduce the convolution

$$\gamma(u) = \int_{R_n} f(u+y)\, f(y)\, dy$$

Finally, let $\mu(S)$ be the volume of S. Then

$$E[Z(S)] = \lambda\, m\, \mu(S) \tag{3.6.1}$$

$$\text{Cov.}\,[Z(S_1), Z(S_2)] = \lambda m\, \mu(S_1 \cap S_2) + \lambda\,(m^2 + \tau^2 - m) \int_{S_1}\int_{S_2} \gamma(u-y)\, du\, dy \tag{3.6.2}$$

It is seen that (2) is of the general form (2.6.2).

If the number of satellites belonging to a center has a Poisson distribution, the above expression simplifies in so far as $\tau^2 = m$. In this case Z is a *stationary compound Poisson process* in the meaning of § 2.6. The corresponding stochastic intensity function is of the kind considered in § 3.2, a moving average process with the constant weight function $m\, f(x)$.

Thomas (1949) has dealt with the degenerate case where all satellites of a particular cluster are contained in an infinitely small region. We then get from (2)

$$\text{Cov. } [Z(S_1), Z(S_2)] = \lambda\, (m^2 + \tau^2)\, \mu(S_1 \cap S_2) \tag{3.6.3}$$

that is a formula of the form (2.6.1), characteristic of orthogonal set functions. $Z(S)$ is an example of a *generalized Poisson process*, cf. § 2.6.

The more general approach in Bartlett (1954) and Thompson (1955) is formally equivalent to the compound Poisson process of § 2.6. The authors quoted give a detailed description of the process. They do not confine their interest to the properties derived from assumptions on $E[\lambda(x)]$ and Cov. $[\lambda(x), \lambda(y)]$. They specify also higher moments or "product densities", $E[\prod_i \lambda(x_i)]$, and derive the corresponding higher moments of the stochastic set function.

The models considered so far give "contagious" distributions of points over R_n. The corresponding frequency distributions of number per unit area is "supernormal". In several applications we would prefer models involving some kind of repulsion between the points ("sub-normal dispersion"). The reason may simply be that the random points actually represent cells of finite size (the random point may be the center or some other point of reference of the cell), so that each point must be surrounded by a certain region where no other point can exist. Cf. the discussion on hemacytometer counts in Turner & Eadie (1957), Hamaker (1958).

A thorough discussion of such processes must involve considerations of time: the random points existing at a certain moment must influence the future course of the intensity function over R_n. We shall here only deal with two very simple models, in which no pair of random points is allowed to have a mutual distance below a certain bound.

Consider a realization of a primary Poisson process of random events. Then exclude every event such that the distance to its nearest neighbour is less than a given positive number R. The remaining points form our model of sub-normal variation. (If two points have a mutual distance $< R$, *both* of them are excluded.) Let λ denote the intensity of the primary process. Suppose that two primary events are located in x and y. The conditional probability that both of them shall be retained in the secondary process is a function, $k(v)$, of the distance $v = |x - y|$:

$$k(v) = \begin{cases} 0 & 0 < v < R \\ \exp[-\lambda\, U(R, R; v)] & R \leq v \end{cases} \tag{3.6.4}$$

with U as in (3.4.7). We find for the secondary process

$$E\,[Z(S)] = \alpha\lambda\mu(S) \tag{3.6.5}$$

where μ denotes volume and α is the probability for a primary point to survive as secondary event:

$$\alpha = \exp\,(-\lambda C_n R^n) \tag{3.6.6}$$

Further

$$\text{Cov.}\,[Z\,(S_1), Z\,(S_2)] = \alpha\lambda\,\mu\,(S_1 \cap S_2) + \int_{S_1}\int_{S_2} \lambda^2 [k\,(|\,x-y\,|) - \alpha^2]\,dx\,dy \tag{3.6.7}$$

Thus the covariance expression is still of the general form (2.6.2). However, it should be observed that the integrand in (7) and the corresponding function in (2) are not cov.f.'s of an underlying stochastic intensity function. We shall use the general term *product densities* for the integrands in (2.6.2) and in the two covariance formulas above.

The process can be changed into a dynamic scheme, by assuming that the primary process proceeds at a uniform speed in the time-interval $0 < t < 1$. The probability that a primary event occurs between t and $t + dt$ in a volume element dx is

$$\lambda\,dx\,dt$$

A primary point P produced at time t is retained in the secondary process if no other *primary* event has occurred within distance R from P *prior to* t. The probability that a primary point survives is

$$\alpha = [1 - \exp(-\lambda\gamma)]/\lambda\gamma \tag{3.6.8}$$

where γ stands for the volume $C_n R^n$ of the sphere of radius R in R_n. The conditional probability that two primary points located in x and y shall both be admitted as secondary points is denoted $k(v)$, where v is the distance $|x-y|$. We have

$$k\,(v) = \begin{cases} 0 & 0 < v < R \\ \dfrac{2\,U\,(1 - e^{-\lambda\gamma}) - 2\,\gamma\,(1 - e^{-\lambda U})}{\lambda^2\gamma U\,(U-\gamma)} & R \leqslant v \end{cases} \tag{3.6.9}$$

where U is written for $U(R, R; v)$, see (3.4.7). The expectation and the covariance of the secondary process can now be found from (5) and (7), if α and $k(v)$ are taken from (8) and (9).

It may instead be prescribed that a primary point P occurring at time t, shall be accepted if no other *secondary* point has occurred before t in the sphere of radius R around P. It seems however that even an attempt to find the probability corresponding to (8) leads to rather formidable mathematics in this case. (Cf. the distinction between type I and type II counter models, see e. g. Pyke 1958.)

3.7. Numerical examples

Some simple numerical examples, which all refer to R_2, can illustrate the exposition of this chapter.

The first example is a random set model of the type (3.4.1) with $\lambda = 6$, $A = 0.17$. An appropriate number of primary points were located over the unit square and a strip of width 0.17 bordering the square. The coordinates were chosen from tables of random sampling numbers. For each point a circle of radius 0.17 was drawn. The ensuing random set (the intersection between the unit square and the union of the circles around the primary points) is shown as a shaded region in fig. 1. The corresponding cor.f. as obtained from (3.4.3) is also shown in the figure.

The same primary points were then used to give a realization of a point process of type (3.2.4) with $A = 0.17$, $B = 1$. This realization was then used as intensity function for a generalized Poisson process. It was found that the

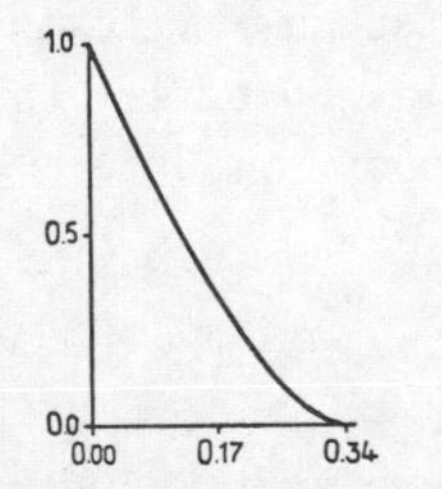

Fig. 1. Model (3.4.1) with λ = 6, A = 0.17
A realization and the correlation function.

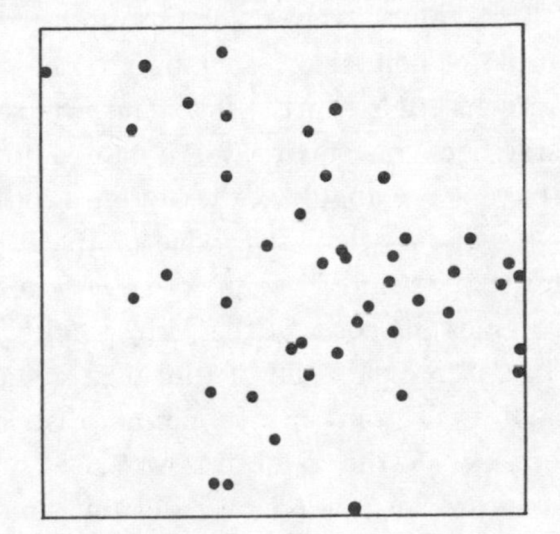

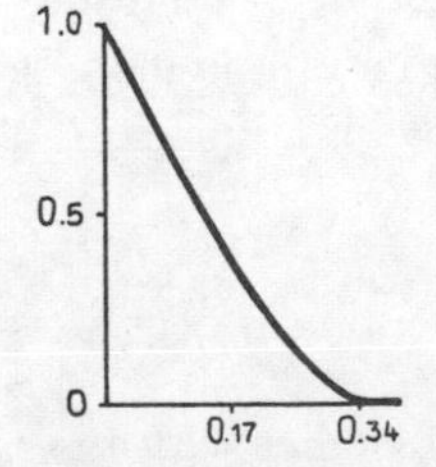

Fig. 2. Generalized Poisson process. Intensity function: (3.2.4.) with A = 0.17, B = 1. A realization of the process and the correlation function of the intensity function.

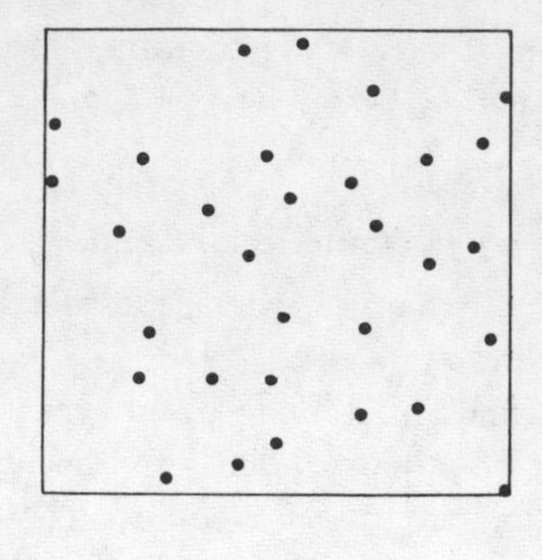

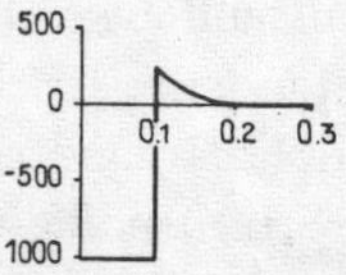

Fig. 3. Subnormal distribution of points. A realisation and the corresponding product density.

highest value attained for the intensity function inside the square was 3. A sample of points corresponding to the generalized Poisson process was therefore constructed in the following manner. A sequence of points $(P_1, \ldots, P_n)$ inside the unit square was obtained from lists of random numbers. A score y_i from a rectangular distribution over the interval (0, 3) was attached to P_i. If y_i was less than the value of the intensity function at P_i, the point was admitted as sample point, otherwise not. The result is shown in fig. 2, which also gives the cor.f. of the corresponding intensity function.

Fig. 2 presents the picture of a "contagious" or supernormal distribution of points. To obtain a subnormal distribution, the last example of § 3.6 was used. For R the value 0.1 was chosen. Thus a primary sequence of points was located at random over the unit square and the bordering strip. A primary point in the square was admitted as secondary point if no primary point had been found earlier within distance 0.1 from the point in question. The procedure was continued until no more secondary points could be obtained. This corresponds to the limiting case $\lambda \to \infty$. The corresponding product density (cf. 3.6.9) is

$$\begin{array}{ll} -1/\gamma^2 & v \leqslant R \\ 2/\gamma U - 1/\gamma^2 & v > R \end{array}$$

It is shown, together with the sample of points obtained, in fig. 3.

It was possible in these three examples to obtain an exact realization over a bounded region in a finite number of operations. For other models, an infinite sequence of steps would be needed. This means that some procedure of approximation must be used. It seems not necessary for the purpose of the present investigation to enter into a discussion of the numerical questions that would be involved in such procedures.

Chapter 4. Some remarks on the topographic variation

4.1. Local and "long-distance" variation

The patterns encountered in studies of the spatial variation often resemble realizations of real-valued stationary stochastic processes. However, this is usually true only as regards local (short-distance) variation. In dealing with variations over long distances we must often introduce an evolutive element in the probabilistic model or add a deterministic smooth trend, as mentioned in § 1.1.

For the applications treated in this paper it is the local variation that is of chief interest. The statistical properties of sampling schemes (and experimental designs) are mostly dependent on the nature of the short distance variation. See the discussion in Matérn (1947, pp. 63 ff.) and Jowett (1955). Of course "local" and "long-distance" are relative concepts. They must be seen in relation to the size of the sampling strata, the experimental blocks, etc.

Furthermore, data (tables, maps) on the spatial variation are often confined to a very limited region, so that inferences cannot be made about the long-distance variation. If a stationary model fits the data over a restricted region, it can often be modified by introducing some additional low frequency waves and still show the same agreement with the data. This means that two processes with the cov.f.'s $\sigma^2 r(u)$ and $A+\sigma^2 r(u)$, respectively, can give the same sort of realizations over a bounded area, if A (>0) is very nearly constant over the range of variation in question. Jowett (1955) has proposed to characterize the variation by a function which does not suffer from this indeterminateness, namely the "*serial variation function*". For a stationary process $z(x)$ in R_n we may accordingly define

$$v(u) = {}^1/_2\, E[|z(x+u)-z(x)|^2] = \sigma^2\{1-\mathrm{Re}[r(u)]\} \tag{4.1.1}$$

It is seen that $v(u)$ does not change if a constant term is added to the cov.f. Incidentally, Langsaeter (1926) used this way of expressing the variation, when dealing with systematic sampling in forest surveys.

Although the variation function has some evident advantages, we shall use the more traditional correlation function also in its empirical form, with a *correlogram* as graphical representation. However, the above-mentioned indeterminateness must be kept in mind (cf. Doob 1953, p. 531). It should also be noted that a knowledge of the correlogram for distances of the same magnitude as the dimensions of the observed region is not of much value, if no additional information is available about the long-distance variation.

In this and the following chapters, only *real-valued processes* will be considered.

4.2. Some data on the spatial variation

In Matérn (1947) a number of correlograms were presented which were based on data from the National Forest Survey of Sweden. The factors studied included some areal distributions (total land, forest land, a particular site class). Here the term *areal distribution* is used as an empirical analogue of the random set model of § 3.5. Correlograms were also presented for the variation in volume of trees. These correlograms were found to exhibit the following general features:

(A) The correlation is *monotonously decreasing* with increasing distance.

(B) The correlation is often nearly isotropic but may show a certain influence of *direction* as well as distance.

(C) The correlograms can be smoothed by curves that have negative derivatives at the origin and are downward convex in the vicinity of the origin.

As to (C) the author used functions of the types $\exp(-av)$ and

$$p \exp(-av) + q \exp(-bv) \tag{4.2.1}$$

with $p, q, a, b > 0$. It is of course possible to get an equally good graduation of the data with several other types of functions that have the properties (A) and (C). Formulas as well as experimental series (cf. Kendall 1946, p. 33) show that an attempt to obtain accurate information about the structure of a stationary process requires an overwhelming amount of data.

In this context it should be pointed out that we are here not concerned with the general inference problem about stationary processes. Our aim is only to get a rough picture of the behaviour of the topographic variation. For the inference problem, see Grenander & Rosenblatt (1956), and literature quoted in that book.

We add now some more correlograms of areal distributions. They all refer to the distribution of land area on maps of the Stockholm region. Thus to each observation point (x) we attach the value $z(x)$ defined as 1 if x is on land, and 0 if x is in water.

The first example is based on a map on the scale 1:250,000, showing a region of 56×68 kilometers around Stockholm (KAK:s bilatlas över Sverige, 1955, blad 23). Twelve equidistant lines with direction E – W were drawn on the map and z was registered for a sequence of points along each line. The distance between two neighbouring points was 1 mm, corresponding to 250 meters in

Table 2. Observed serial correlations (r_k) of the distribution of land area in the Stockholm region.

Lag (k)	Series 1	Series 2	Series 3	Series 4
	Map on the scale 1:250,000	Map on the scale 1 : 50,000		
	East-west	Four directions	East-west	North-south
0	1.000	1.000	1.000	1.000
1	0.736	0.924	0.546	0.605
2	0.601	0.868	0.463	0.485
3	0.537	0.834	0.395	0.423
4	0.491	0.803	0.332	0.393
5	0.442	0.767	0.294	0.331
6	0.397	0.748	0.284	0.303
7	0.361	0.721	0.261	0.277
8	0.328	0.708	0.203	0.245
9	0.316	0.690	0.176	0.207
10	0.308	0.656	0.168	0.205
11	0.285			
12	0.255			
15		0.606		
20		0.572		
Lag 1 corresponds to the interval:				
on map	1 mm	1 mm	10 mm	10 mm
in the field	250 m	50 m	500 m	500 m
No. point pairs for a correlation	3,120	1,945—3,259	2,091—2,550	2,091—2,550

the field. When the serial correlation, r_k with lag k mm ($k \leq 12$), was computed the 260 first points (counted from the west) on all lines were used as the fixed series; the corresponding 260 points k steps to the east formed the other series. (The total number of points on each line was 272). Thus each correlation is based on $12 \cdot 260 = 3120$ pairs of points. The results are shown in table 2 and figure 4.

Table 2 and fig. 4 also contain three other series of correlations, which are all based on a map (scale 1:50,000) over a part of the region shown on the previous map (Topographic map of Sweden, Stockholm SE, Generalstabens litografiska anstalt 1955). The area covered is 25 × 25 kilometers.

From each one of 18 points systematically located in the map, four rays of length 50 mm were drawn in directions NE, N, NW, and W. On each ray $z(x)$ was registered for points at distance 1 mm (50 meters) apart. Since a few lines cut the edge of the map, less than 50 points were surveyed in some cases. All pairs on a line with mutual distance k mm were used for the calculation of a correlation r_k. Thus the correlation coefficients are based on a varying number of point pairs (e.g. 3259 for r_0, and 1945 for r_{20}). The results are shown as series 2 in table 2 and figure 4.

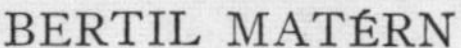

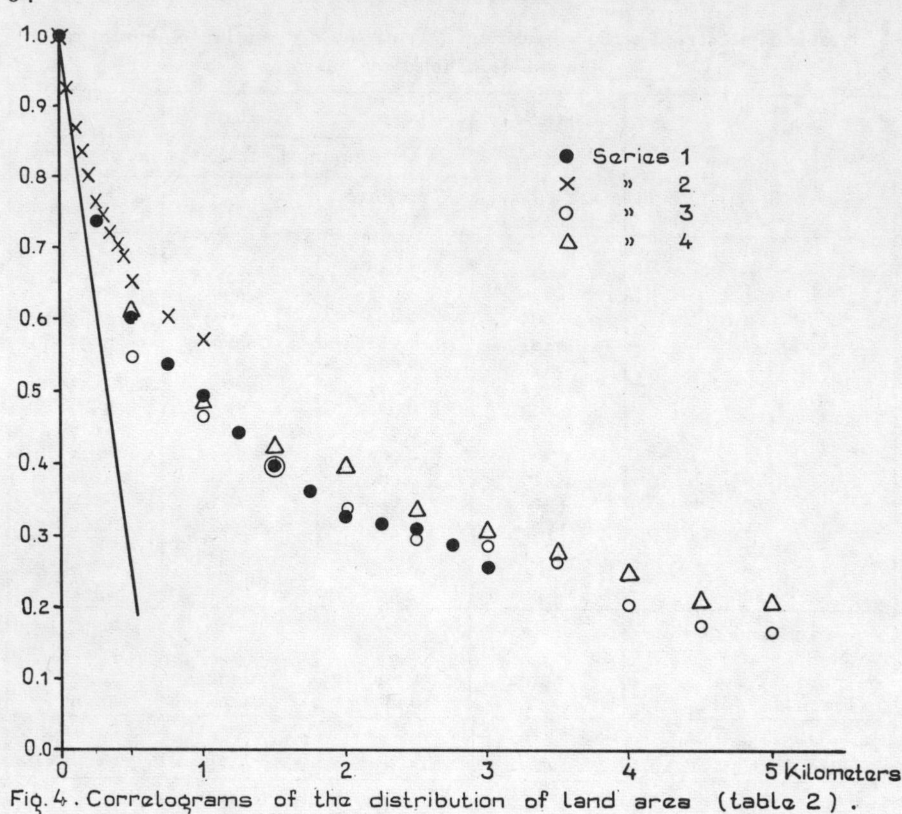

Fig. 4. Correlograms of the distribution of land area (table 2).

The series 3 and 4 are based on parallel lines running across the map in direction east-west (series 3) and north-south (series 4). Points 10 mm (500 meters) apart were observed. Also in these cases all possible pairs of points on a line were utilized when computing the correlation coefficients.

The estimated derivative at the origin for the correlogram of series 2 is also entered in fig. 4. It was found that the rays surveyed had a total length of 326.2 cm and a total of 88 intersections with the shore-line. If the centimeter unit is used, we find for the derivative of the empiric covariance function according to (3.5.7)

$$c'(0) = -\theta/2 = -{}^1/_2(88/362.2)$$

Further the water area (sea and lakes) comprised 23.18 per cent of the points enumerated on the lines. Hence

$$r'(0) = c'(0)/\ 0.2318(1-0.2318) = -0.757$$

Incidentally, it may be noted that (when the rays cover the variation in four directions) we can fairly well estimate the length of the shore-line in centimeter per sq. centimeter as $\theta\pi/2 = 0.424$. This corresponds to 848 meters per sq. kilometer in the field. (See formula 3.5.8.)

The properties (A) and (B) seem to be rather generally recognized, see references in Matérn (1947). (As to the early discussion of the variation in terms of correlation functions, the author wants to add to these references the works of Mahalanobis 1944 and Nair 1944.) Also (C) seems to be recognized by some authors as a common trait of the topographic correlation (see Quenouille 1949, Jowett 1955), although, as pointed out by Whittle (1954), the exponential correlation mentioned earlier in this section cannot claim any divine right.

Unfortunately, not many correlograms of the topographic variation can be obtained from published data. It may be noted, however, that the observations of Williams (1952, 1956) as well as those of Zubrzycki (1957) and Whittle (1954) agree with (A) and (B). However one of Whittle's correlograms (p. 445) disagrees with (C).

Now it is clear from § 3.5 that the cor. f.'s of areal distributions must have the property (C). Also the variation of many other factors, e. g. soil fertility, may be influenced by discontinuities in the underlying areal distributions.

Property (C) is a question of the behaviour of the variation for very short distances. Some special circumstances that are of interest in this context will be discussed in the subsequent sections. The influence of errors of measurement will be dealt with in 4.3. In 4.4 we shall briefly comment on the "integration" of the fertility in the neighbourhood that can be thought to be represented in the growth of a plant. Some comments on the competition between plants and its effect on the short distance correlation will be presented in 4.5.

Indirect observations on the covariance structure of the topographic variation are given by authors reporting on the efficiency of various designs of field experiments and areal sampling. Such observations rather unanimously confirm (A), see Matérn (1947, p. 22). However, the possibility of a *periodicity* in the topographic variation has been discussed, especially in connection with systematic sampling. One well-known example by Finney (1950) will be briefly reviewed in the concluding section of this chapter.

As to early reports on the topographic variation, the following contributions should be added to those mentioned in Matérn (1947): Harris (1915), Smith (1938); see also the discussion and the literature quoted by Cochran (1953, pp. 176 ff.) and Milne (1959).

4.3. Errors of observation

In several cases it is appropriate to think of the *observed process* (z) as the sum of the *true process* (z_1) and a *superimposed error* (z_0). It should be clear that the observed process is known only in a finite number of points that not necessarily form a regular pattern.

If z_0 has the nature of an *error of measurement*, it can be regarded as a "chaotic" process, independent of z_1, with cor.f. (2.2.1). Then, the cor.f. of z gets a discontinuity at the origin, see (2.2.2).

However, the same type of cor.f. can also result from other causes. A factor (e.g. growth) connected with plants located in discrete points, may be considered as composed of two parts, one with a continuous variation expressing the influence of the environment, and the other of the chaotic type representing the effect of the genetic structure of the plant (cf. Whittle 1954, p. 445).

Returning to the observational errors, we shall now shortly consider a type of inaccuracy which can be called "displacement error". It can be expressed in the formula

$$z(x) = z_1(x_1)$$

where x_1 is a point in the neighbourhood of x, which happens to be selected, when we attempt to locate x. It is realistic in some cases to assume that all vectors $(x_1 - x)$ have a common frequency function $q(y)$ and that all displacement errors are independent. For the cor.f. of the observed process we then have

$$r(u) = \int \int q(y_1)\, q(y_2)\, r_1(u + y_1 - y_2)\, dy_1\, dy_2 \tag{4.3.1}$$

where r_1 is the cor.f. of the true process. The formula is valid for $u \neq 0$. It is clear from (1) that also in this case the cor. f. of the observed process has a discontinuity at the origin. For $u > 0$, $r(u)$ is a weighted average of the values of r_1 in some neighbourhood of u. The properties of this type of cor.f.'s will be discussed further in the next section.

We shall also consider *rounding off* errors. Let the observed $z(x)$ be defined as a multiple, nh, of the length, h, of the rounding-off-interval, with n determined from

$$nh \leqslant z_1(x) < n(h+1)$$

We thus consider rounding off to the *nearest lower* multiple of h. If the rounding off is to the *nearest* multiple of h, the covariance structure is the same. Now, if the distribution of each particular variable $z_1(x)$ is such that Sheppard's corrections apply, approximately

$$c(0) = c_1(0) + h^2/12 \tag{4.3.2}$$

When the corresponding conditions for the two-variate distribution of $[z_1(x), z_1(x+u)]$ are valid, we similarly have (cf. Wold 1934)

$$c(u) = c_1(u) \tag{4.3.3}$$

A strict derivation of these results is given by Grenander & Rosenblatt (1956, pp. 55 ff.) in the case of a discrete-parameter normal (Gaussian) process.

In the case of a continuous-parameter process, it must be noted that (3) cannot be used when $c_1(0) - c_1(u)$ is of the same or lower order of magnitude than h^2. It is for example intuitively clear that if $c_1(u)$ is continuous at the origin, the same must hold true of $c(u)$.

We shall now indicate how the properties of $c(u)$ for small u can be found. Only the case where c_1 is continuous at the origin, and h^2 is small compared to $c_1(0)$, will be treated. Consider the two random variables $z_1(x)$ and $z_1(x+u)$. Assume for the present only that the correlation of $z_1(x)$ and $z_1(x+u)$ is not in the vicinity of -1. Thus the sum of the variables has a distribution which is not concentrated in an interval of the magnitude of h. Introduce two auxiliary random variables by

$$\alpha = \frac{z_1(x+u) + z_1(x)}{h} \qquad \beta = \frac{z_1(x+u) - z_1(x)}{h}$$

Suppose that the corresponding distributions are of the continuous type. Let $p(t)$ denote the frequency function of β, and let $q(s|t)$ denote the conditional frequency function of α given $\beta = t$ (see Cramér 1945, § 21.4). Consider also the random variable

$$\tau = [z_1(x+u)/h] - [z_1(x)/h]$$

where generally $[a]$ is the largest integer $\leqslant a$. Let us further use the notation

$$p_n = P(\tau = n)$$

Now, write β as $n + t$ with $0 < t < 1$. The conditional probability of the event $\tau = n$ is seen to be

$$\sum_{\nu=-\infty}^{\infty} \int_{t}^{2-t} q(n + 2\nu + s \mid n + t)\, ds$$

Thus we integrate over a part of length $2(1-t)$ of every interval of length 2. The above sum can therefore be approximated simply as $1 - t$. Adding the corresponding integral for the case $-1 < t < 0$, approximately

$$p_n = \int_{-1}^{1} (1 - |t|)\, p(n+t)\, dt$$

If $E(\beta^2)$ is large and if the usual assumptions are made about high contact at the end-points of the range of $p(t)$, (3) follows from the above formula. If on

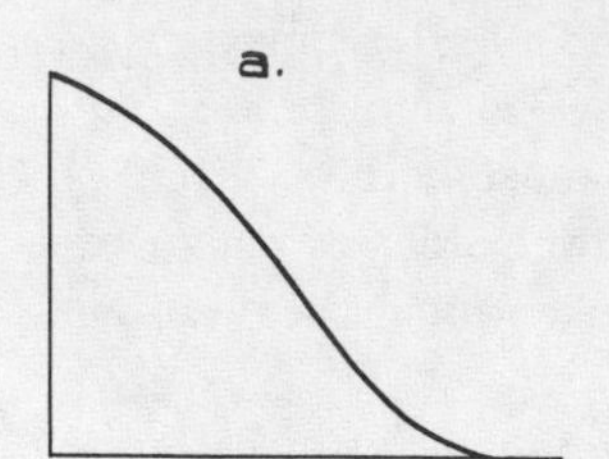

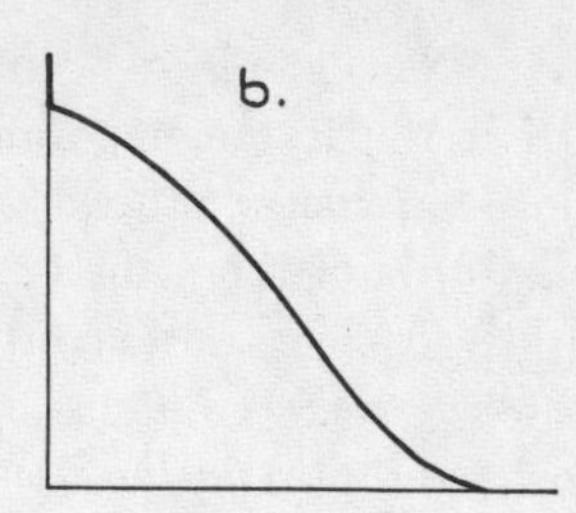

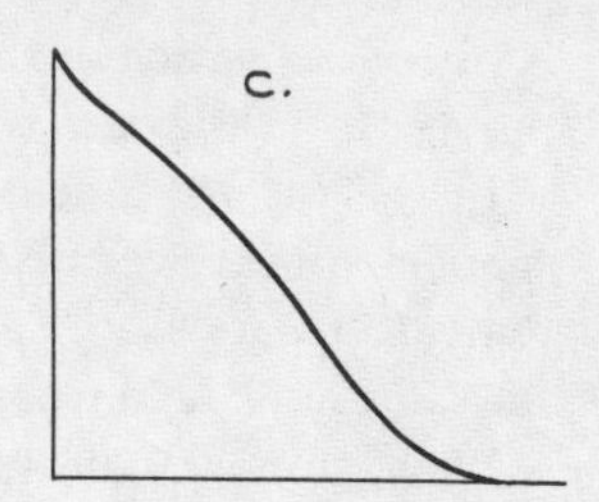

Fig. 5. Original covariance function of a process (a); of the process with superimposed errors of measurement (b); of the process with superimposed rounding off errors (c).

the contrary $E(\beta^2)$ is small, we need only consider p_{-1}, p_0, and p_1 in seeking an approximation. In this case it can be seen that

$$E(\tau^2) \approx \int_{-1}^{1} (1 - |t|) [p(1+t) + p(-1+t)] dt$$

$$\approx \int_{-\infty}^{\infty} |t| \, p(t) \, dt = E(|\beta|)$$

Now we must have

$$E(|\beta|) = k_u \sqrt{E(\beta^2)}$$

where k_u is ≤ 1 (see Cramér 1945, formula 15.4.6). In the case of a normal distribution k_u equals $\sqrt{2/\pi}$. Thus, the following approximation is valid

$$c(u) = c(0) - hk_u \sqrt{\frac{c_1(0) - c_1(u)}{2}} \tag{4.3.4}$$

if u is such that

$$c_1(0) - c_1(u) = o(h^2)$$

This shows that $c(u)$ is continuous at the origin. However, it is also seen that the graph of $c(u)$ must show a cusp at the origin even if $c_1(u)$ has zero derivatives for $u=0$. This follows from theorems on the behaviour of a ch.f. in the neighbourhood of the origin, see Cramér (1937, p. 26). It is also intuitively clear from § 3.5, since $z(x)$ is discrete-valued.

The difference in influence on the covariance structure between the various types of errors is shown in fig. 5. (The figure refers to the isotropic case. It can also be thought of as giving the covariance in one particular direction in the general case.)

In many cases several types of errors of observation are in action at the same time. The resulting cov.f. may then be a hybrid between the types (b) and (c) of fig. 5.

Incidentally, we can also draw another conclusion from (4), which we formulate for an isotropic process in R_2. Suppose that each realization of $z_1(x)$ is continuous and that the expected length of *level curves* (drawn with a certain interval of altitude, h) is finite in any bounded area. Then the derivative $c'(0)$ of the rounded off process must have a finite negative value, which implies that $c'_1(0) = 0$. Assuming a Gaussian process and applying (3.5.5) to $z(x)/h$, we find for the length (per unit area) of the level curves of z_1 the expected value

$$\frac{1}{h}\sqrt{\frac{\pi|c_1''(0)|}{2}}$$

4.4. Local integration

Let c_1 be the cov.f. of the stochastic process z_1. Assume that the spectrum is absolutely continuous with spectral density $f(x)$. A new process if formed by the relation

$$z(x) = \int q(x-y)\, z_1(y)\, dy$$

where the integration takes place over the range of $q(x-y)$ which is assumed to be some neighbourhood of x. We shall only deal with the case where q is a fixed function. This type of "local integration" can represent for example the influence of the environment on a plant located in x, see Whittle (1954, p. 445).

Denote by φ the Fourier transform of q. It is seen that the spectral density of z is

$$\text{const. } |\varphi(x)|^2 f(x) \tag{4.4.1}$$

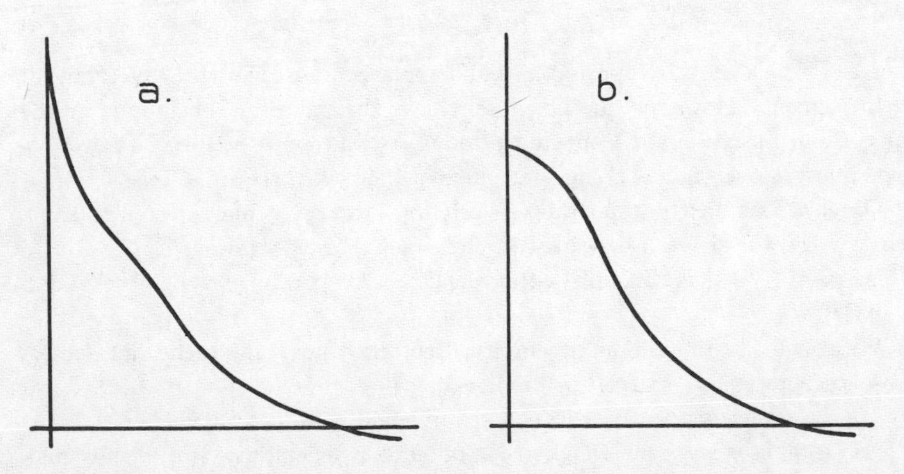

Fig. 6. Covariance function of the orginal process (a); of the process transformed by local integration (b).

Table 3. Empirical covariances of the integrated intensity function $z(x)$ (see text) for the distribution of coniferous plants.

	Field 37: I					Field 37: II			
	Values of $c(i, j)$					Values of $c(i, j)$			
	$j=0$	$j=1$	$j=2$	$j=3$		$j=0$	$j=1$	$j=2$	$j=3$
$i=-3$		—0.71	0.03		$i=-3$		1.01	—0.04	
$i=-2$		1.08	1.63	1.18	$i=-2$		1.16	0.49	1.07
$i=-1$		2.51	2.11	1.33	$i=-1$		1.31	0.66	0.96
$i=0$	4.91	2.62	2.31	2.20	$i=0$	3.63	2.02	0.81	0.86
$i=1$	2.65	1.77	0.63	0.47	$i=1$	2.69	1.90	1.21	0.75
$i=2$	1.36	1.14	—0.17	0.72	$i=2$	1.59	1.45	1.73	0.43
$i=3$	—0.30	—1.55	0.26		$i=3$	1.20	1.47	1.29	

Distance	Average of $c(i, j)$	Distance	Average of $c(i, j)$
0	4.91	0	3.63
1	2.63	1	2.35
$\sqrt{2}$	2.14	$\sqrt{2}$	1.60
2	1.84	2	1.20
$\sqrt{5}$	1.24	$\sqrt{5}$	1.12
$\sqrt{8}$	0.73	$\sqrt{8}$	1.11
3	0.95	3	1.03
$\sqrt{10}$	—0.11	$\sqrt{10}$	1.05
$\sqrt{13}$	0.55	$\sqrt{13}$	0.68

Unit of distance: 2.5 cm in the map, corresponding to 2.5 meters in the field

(Doob 1953, p. 535). The cov.f. of z is

$$c(u) = \int\int q(y_1)\, q(y_2)\, c_1(u + y_1 - y_2)\, dy_1\, dy_2 \tag{4.4.2}$$

i.e. a formula of the type (4.3.1) valid for a process distorted by errors of displacement. However, in the present case the cov.f. is continuous at the origin, even if $c_1(u)$ is discontinuous for $u=0$, and (2) is valid also for $u=0$. The effect is one of smoothing the course of the cov.f., see fig. 6.

Observations on the topographic variation are very often referred to some basic cells of positive area (squares, circles, etc.). The formulas apply also in this case if $q(x-y)$ is constant within the cell corresponding to the observation point x.

We conclude this section by empirical correlograms, where the (indirectly) observed process is of the integrated type.

The correlograms are based on maps of two experimental fields (37: I and 37: II) established by the Department of Forest Regeneration of the Swedish Forest Research Institute. On each map, a square network of circular sample plots was surveyed. The radius of each circle was 0.7 cm (corresponding to 0.7

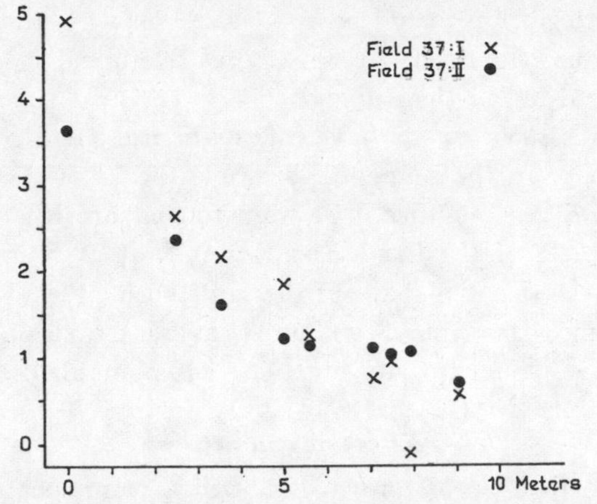

Fig. 7. Empirical average covariances for the intensity function of the distribution of coniferous seedlings. (table 3)

meters in the field). The distance between the centers of the circles was 2.5 cm. The network comprised $14 \times 14 = 196$ sample plots in each field. For each plot the number of coniferous seedlings was observed.

Although the amount of data is rather limited, we may try to find the type of intensity function (cf. 2.6 and 3.6) that may produce a distribution of plants of the kind encountered. Then, let $z(x)$ designate the integral of the intensity function over a circle of radius 0.7 around x.

In field 37: I the number of seedlings per plot was $800/196 = 4.08$. The variance of seedlings per plot was

$$(5018 - 800^2/196)/195 = 8.99$$

Formula (2.6.2) shows that this corresponds to the following variance of the integrated intensity function $z(x)$

$$c(0, 0) = 8.99 - 4.08 = 4.91$$

This value and the corresponding values of $c(i, j)$ (covariance between two circles with the following distances between the centers: i steps in x-direction, and j steps in y-direction) are entered in table 3, which also contains corresponding information from field 37: II. The average number of plants per plot ($410/196 = 2.09$) was lower than in field 37: I.

It has already been remarked that the number of observations is small; e.g. the covariance $c(3, 2)$ is estimated from only 132 pairs of points. Although there is some evidence of a directional effect on the variation, the covariances in different directions have been averaged for each distance. These averages are found in the lower half of table 3 and they are also shown in fig. 7.

The effect of integration should be appreciable only in the case $c(0, 0)$. The corresponding value $c_1(0, 0)$ of the cov.f. of the original intensity function should be somewhat higher than $c(0,0)$.

Occasionally it is sufficient to have information about a process obtained by local integration. In the numerical example the knowledge of the above covariance function c may suffice if we want to compare different schemes for locating sample plots of the fixed size. However, if we want to study the effect of varying the size and shape of the sampling unit, it will be necessary to delve deeper into the "micro-structure", and information about the covariance, c_1, of the "point-intensity-function" will be needed.

4.5. Effect of competition

Let $z(x)$ denote the size (or growth, etc.) of a plant located in x. If the difference $u = x - y$ between the coordinates of two plants is small, it may be expected that the correlation between $z(x)$ and $z(y)$ should be negative owing to the competition between the plants. However, published observations on the variation of the yield of agricultural fields and forest stands do seldom exhibit traits of the kind that would be induced by this competition (exceptions are Hudson 1941, Johnson & Hixon 1952). It is possible that the correlation in soil properties between neighbouring points is so strong that the effect of the competition is obscured. Another possible explanation is that obser-

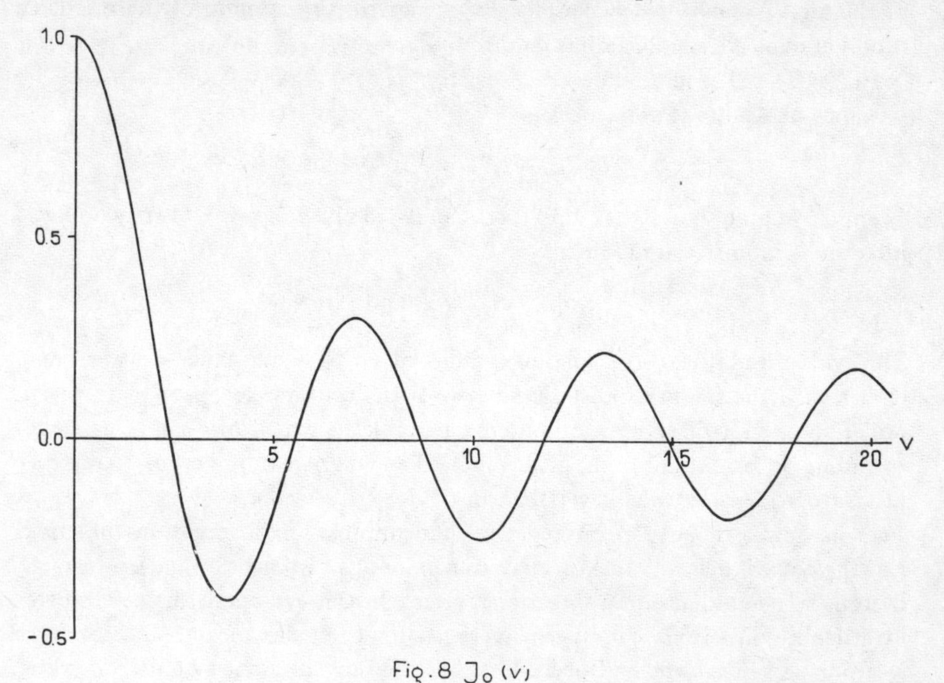

Fig. 8 $J_0(v)$

vations seldom are made at very short distances. Also, as pointed out in the preceding section, observations often refer to areas and not to points. These areas may be large compared to the distances over which the competition is effective.

The following process is of interest for certain applications in forest surveys. To each point x is attached the value $z(x)$ of the height of that part of a stem which is perpendicularly above x. If $z(x)$ is integrated over an area containing one tree, the integral thus gives the volume of this tree. The cor.f. $r(u)$ of $z(x)$ must be continuous at the origin. The effect of competition should manifest itself in low values of $r(u)$ at some distance from the origin (this distance depends on the size of the trees), and the effect should be repeated as a sort of damped oscillations. In the isotropic case in R_2 it is clear that the extreme case of "periodicity" results in a cor.f. of the type

$$\Lambda_0(av) = J_0(av)$$

see (2.3.7). This function, with $a = 1$, is shown in fig. 8.

It has generally not been considered necessary to take into account the effect in the following applications. Although the short-distance variation here is of interest, the effect of competition can be considered as belonging to a "micro-state" with which we need not be concerned.

4.6. The occurrence of periodicities in the topographic variation

The possible existence of a more or less strictly periodic component of the topographic variation has been mentioned by several authors, especially in discussions of systematic sampling.

In the extreme case, a realization of the process $z(x_1, x_2)$, where x_1 and x_2 are the rectangular coordinates of a point in R_2, would contain a component such as

$$\cos(Ax_1 + B) \tag{4.6.1}$$

In (1) the wave-front moves perpendicular to the x_1-axis. The cyclic pattern will here be conspicuous if observations are taken along paths parallel to the x_1-axis, e.g. in observations of the one-dimensional process

$$Z(x_1) = \int_a^b z(x_1, x_2)\, dx_2 \tag{4.6.2}$$

An instance of periodicities which has been extensively debated was reported by Finney (1950). The data studied by Finney pertained to the average volume per acre obtained from 292 strips in a survey of a forest at Dehra Dun, India. Systematic samples with varying sampling intensities were taken from these data, and the sampling variances of the corresponding means were computed. When these variances were plotted as a function of the sampling ratio 1 : p, some striking irregularities were found. A marked peak was observed for $p=17$, which indicates a possible periodicity with a cycle comprising seventeen strips.

It has been suggested that irregularities of the same magnitude might be obtained if purely random series are analysed in a similar way (Matérn 1953). However, it is found in the Dehra Dun example that the means of the seventeen systematic samples show a wave-like pattern, see curve II in fig. 10.

An effect of this kind might be expected to appear in many types of autoregressive series. It may suffice to consider a simple Markoff chain. Hence assume that in the series

$$z_1, z_2, \ldots, z_n \tag{4.6.3}$$

the covariance of z_i and z_j is $\exp(-hv)$, where $v=|i-j|$. To avoid unnecessary complications suppose further that $n=pq$, where p and q are integers. The means of p systematic samples from (3) with sampling ratio 1 : p are

$$\bar{z}_i = (1/q) \sum_{j=0}^{q-1} z_{i+pj} \qquad (i=1, 2, \ldots, p) \tag{4.6.4}$$

Then with $v=|i-j|$

$$\text{Cov.}\,(\bar{z}_i, \bar{z}_j) = \frac{\exp(-hv) + \exp[-h(p-v)]}{q - q\exp(-hp)} - \frac{\cosh(hv)}{2q^2} \cdot \frac{1-\exp(-hn)}{\sinh^2(hp/2)} \tag{4.6.5}$$

When p and q are large the covariance is approximately proportional to

$$\exp(-hv) + \exp[-h(p-v)]$$

Hence the random variables (4) are approximately cyclically correlated.

To obtain a numerical illustration we consider a series of 300 values computed in the following way

$$x_t = 100(5 + y_t)$$

where for $t>0$

$$y_t = 0.96\, y_{t-1} + 0.28\, \varepsilon_t$$

The numbers ε_t and y_0 were taken from Wold's table of Random normal deviates (Wold 1948). Thus a stationary series with cov.f. $(0.96)^t$ was obtained.

All possible systematic samples of intensity 1 : p with $p=2, 3, \ldots, 30$ were selected from the ensuing series of 300 z-values. A variance per sample unit was then computed from the p sample means corresponding to the ratio 1 : p. The variances are found in table 4. They are also shown in fig. 9.

Table 4. Variance per sampling unit in systematic samples from an autoregressive series. Sampling ratio: $1/p$.

p	Variance per unit
2	12
3	91
4	613
5	219
6	144
7	349
8	913
9	813
10	719
11	672
12	644
13	558
14	336
15	955
16	3,100
17	715
18	1,585
19	1,530
20	1,126
21	604
22	2,317
23	2,686
24	1,060
25	1,478
26	947
27	1,223
28	1,834
29	1,633
30	1,521

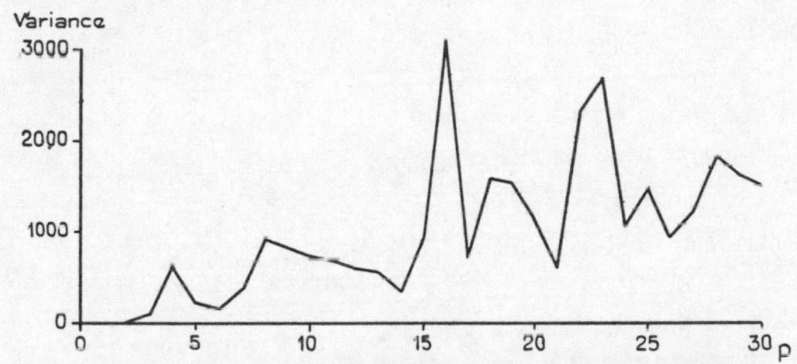

Fig. 9. Variance per sampling unit in systematic samples from an autoregressive series. Sampling ratio: $\frac{1}{p}$. (Data from table 4.)

It should be remarked that this is not the technique used by Finncy in his computations. Finney's method is described in his earlier paper (1948).

The highest peak in fig. 9 is found for $p=16$. The variance per sample point is 4.3 times larger than the value corresponding to $p=17$. Thus the interval 16 would be rather unfavourable in systematic sampling of this particular realization of the process. The means of the 16 systematic samples with $p=16$ are shown in fig. 10, together with the means of the systematic samples with ratio 1:17 from the Dehra Dun data. In both cases we have a smooth curve not very different from a sine wave. Since the 16 means from the autoregressive series are linked together in a circle with close correlation between neighbours, some pattern of that kind should also be expected, cf. (5).

In conclusion, the rather manifest periodicity in the sample means of the autoregressive series cannot be attributed to any underlying cycle-producing mechanism. Its high amplitude can be explained as a result of the selective procedure used to choose the particular wave-length studied. (Certain more complicated schemes of the autoregressive or moving-average type are known to induce periodicities with wave-lengths and amplitudes that can be predicted from the underlying mechanism, see Slutzky 1937 and Wold 1938.)

What are the implications as to the usefulness of systematic sampling in obtaining information about the mean value of a series such as the one considered here?

When an empiric series is analysed in the way illustrated by fig. 9, one must find some values of p for which the systematic sampling gives comparatively poor efficiency. Therefore, if systematic sampling is practised, occasionally values of p will be used, that are not very well suited to the particular series.

However, it can be argued that something of this nature may happen also when a random selection of sampling units is applied. To take an example, consider all possible ways of selecting a sample of 5 units from a population of 20 elements, numbered 1, 2, . . . , 20. A set of 15,504 different combinations is then obtained. This set can be partitioned into e.g. 272 subsets each with 57 members. It seems safe to assume that for any population of 20 units, one or two of the 272 subsets will be rather unsuited for use in sampling. When unrestricted random sampling is applied, the sample actually selected will sometimes belong to one of these subsets which are especially unfavourable for the population in question.

Although the analogy is far from perfect, this may suffice as an illustration of the simple fact that when an actual sample is chosen (or considered as chosen) from a small number of combinations of units, the precision of the sample may now and then come out as rather low, on other occasions as extremely high. This cannot be considered as a very serious argument against systematic sampling.

However, the possibility of mechanisms producing strong periodicities must be excluded if systematic sampling is to be advised; or the wave-lengths should be known in advance, so that an unfavourable sampling interval could be avoided.

If cyclic patterns were common in the topographic variation, they would represent a serious problem in forestry, since systematic sampling is extensively practised in this field. For this reason it is of interest to review briefly the opinion expressed by A. Milne (1959) after a discussion of the problem and a close scrutiny of the Dehra Dun data.

Milne's article disclosed that the 292 strips in the survey actually consisted of two distinct sets of equidistant strips (strips 1—187 and 188—292 with an unsurveyed portion corresponding to 24 strips between no. 187 and no. 188). The same periodic pattern was found in both sets. There was, however,

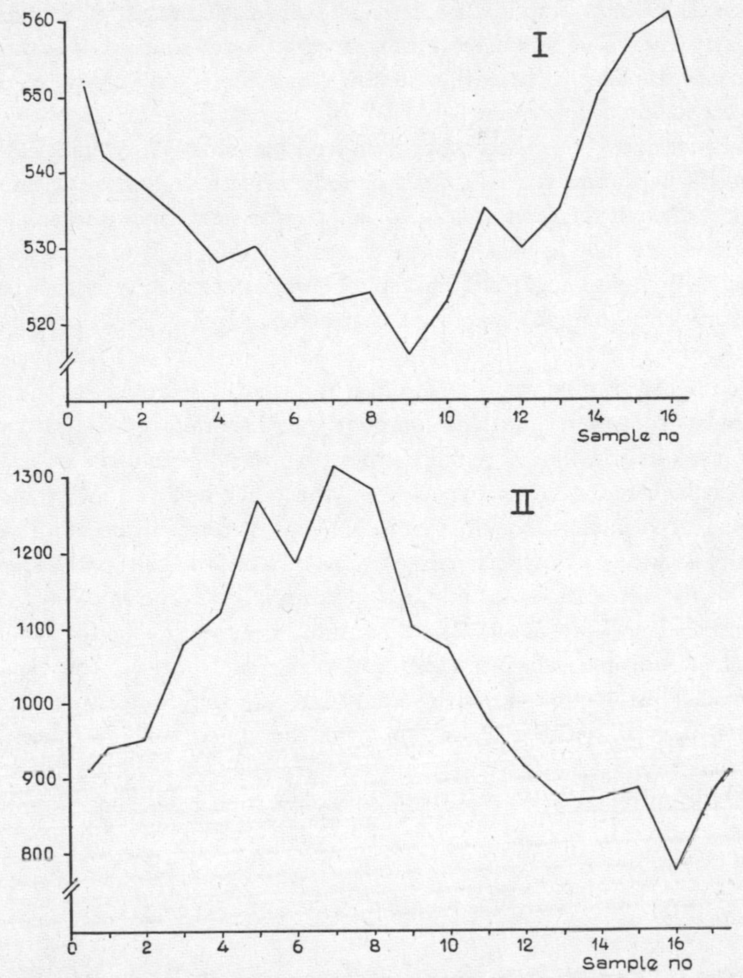

Fig. 10 Means of systematic samples from an autoregressive series (I) and from observations of the cubic volume (II).

a displacement of phase equal to half a "period". These findings and other considerations gave Milne "the not unreasonable impression that the periodicity does not in reality exist in Dehra Dun but is a result of the mechanism of the enumeration".

Milne also studied 20 complete enumerations (yields of agricultural crops, horticultural and orchard crops; number of larvae, eggs or adults of several kinds of beetles) and discovered no sign of periodicity. Nor could he find "reasonable grounds for expecting spatial periodicity anywhere on this earth except where man himself, either directly or indirectly, has imposed periodic conditions sufficiently accurate to override the natural environmental irregularity". Milne also was of the opinion that "man-made spatial periodicity will nearly always be suspected either from external signs or past history". The surface drainage of irrigation channels and the equal spacing of planted trees were mentioned as examples.

Milne's formulation must probably be considered as an overstatement. It is not difficult to find spatial "quasi-periodicity": the reflection of the yearly cycle in sedimentary rocks and soils, the regular patterns in organic tissues (cf. Ladell 1959), not to speak of sea-waves and the effect of their action on the shore. However, regarding the spatial variation that is encountered in forest surveys, it may be concluded that no clear case of periodicity has been reported.

If a cor.f. of a stationary process in R_2 is isotropic or "elliptic" with moderate excentricity (see end of § 2.4), it can only contain a damped oscillatory component, as stated in the preceding section, see fig. 8. The same holds true a fortiori in higher dimensional spaces. On the other hand, strict periodicities can be induced in cor.f.'s in R_1 by mechanisms of competition etc. that only give damped waves in higher spaces; examples are found in 3.5. A short discussion of the implications for systematic sampling is found in 6.2.

Realizations of a process in R_2 can contain linear waves such as (1) even if the cor.f. is isotropic, which is clear from the model (3.1.2). The experience reviewed in this chapter, however, would indicate that we can exclude such patologic cases in seeking abstractions of the topographic variation. This means that rather far-reaching qualities of ergodic nature might be assumed. The empiric correlations of a realization would then have "nearly isotropic" properties.

Chapter 5. On the efficiency of some methods of locating sample points in R_2

5.1. Introduction

This chapter is concerned with the estimation of an integral

$$\int_Q z(x)\, dx$$

by means of the values attained by $z(x)$ in a certain subset, the "sample", of the set Q. As an illustration of the use of the theory of stationary processes in such contexts, we shall at some length discuss one particular example, namely the case when Q is a region in R_2 and the sample consists of a finite number of points in Q.

We shall mainly deal with the problem of estimating the average (cf. 2.5)

$$z(Q) = \int_Q z(x)\, dx \Big/ \int_Q dx$$

by the statistic

$$\bar{z} = (1/n) \sum_{i=1}^{n} z(x_i)$$

where $(x_1, x_2, \ldots, x_n)$ are the *sample points*.

The reason for this restriction to the unweighted mean is the notion that the use of any other estimator is advisable only in cases where rather detailed prior information about the structure of the process is at hand. Our interest here is not to find the optimum estimation procedure for particular processes; instead we study how the "robust" estimate $\bar{z}$ performs under different conditions.

We shall speak of a *sample design (plan, scheme)* to denote a set of rules for choosing the sample points. These rules may involve a random experiment, but not necessarily so. The main object will be to compare the performance of different sample designs.

A good sample scheme should lead to a low value of the sampling error, $\bar{z} - z(Q)$, and it should further be reasonably unexpensive to apply in practice. We shall in this chapter give some comments on the aspect of *costs*, but the emphasis will be on the *precision* of the estimates.

One way of obtaining an idea of the precision is, of course, to carry out sampling experiments in actual populations of the type appearing in a special field of application. When the sampling is of the random type, the precision can be expressed in simple estimates of the variation existing in the population under survey. If the random element is small or completely lacking (for example in so-called systematic sampling), the collection of empirical evidence about the precision is a very painstaking work. Extensive research of this kind, however, has been made in the case of the "line surveys" used in forestry; the first large-scale such investigations seem to be those of Langsaeter (1932) and Östlind (1932).

Another approach is the one of W. G. & L. H. Madow (1944). These authors studied the average outcome of a (more or less restricted) random selection in the case of an analytically defined function (in R_1).

We shall instead study the average performance of a design in a family of functions. To characterize the performance we shall use the variance

$$E\{[\bar{z} - z(Q)]^2\} \tag{5.1.1}$$

where the expectation is taken over the family of functions and also over the possible outcomes of the random procedure that may form part of the sample plan. This is the approach of Cochran (1946).

The family of functions will be restricted to *the set of realizations of a stationary and isotropic process that has a decreasing correlation function*. In a particular field of application it might of course be useful to consider some more special type of process. However, existing evidence about the topographic variation (see Ch. 4) indicates that a first comparison of sample plans should be confined to the above case. The restriction to isotropic processes may be considered a severe limitation. As indicated in § 5.2 below, however, it is very easy to form conclusions about the case of "*elliptic correlation*" from a study of the isotropic processes. If the isotropic case is considered of prime interest, the elliptic case deserves the second place.

The variance (1) depends not only on the structure of the process and the sample scheme adopted. It is also dependent upon the size and shape of the region Q. If the number of sample points is small, the geometry of Q produces a "*border effect*" that may be rather complicated. It seems however reasonable that we neglect this effect in seeking a first characterization of the sample designs. We shall therefore concentrate on the limiting case where the average number of sample points per unit area is fixed, whereas the region Q extends to infinity in all directions. We shall then mainly consider the *variance per sample point*, defined as

$$\lim_{n\to\infty} n \cdot E\{[\bar{z}_n - z(Q_n)]^2\} \tag{5.1.2}$$

where Q_n can be specified for example as a square of area nA, A being a fixed number; $\bar{z}_n$ is the mean pertaining to the sample points in Q_n.

The results may be applied also in the case where the sampling units are plots of positive area. Each observed $z(x)$ should then be considered as the integral of some primary process z_1, the integration being performed over a certain plot with x as center. If not only z_1 but also z shall be isotropic, this sample plot should be circular, or its orientation should be chosen at random. However (except for distances of the order of magnitude of the diameter of the plot) the isotropic properties of z_1 are practically retained even with other types of sample plots. It may further be necessary to include in the covariance of $z(x_1)$ and $z(x_2)$ a component proportional to the area of the intersection between sample plots centered in x_1 and x_2, respectively; see the discussion attached to (2.6.2) and (3.6.7). A purely "chaotic" component should be added to describe the influence of errors of observation. Thus it can be assumed that the cov.f. is a sum of two components

$$c = c_1 + c_2 \tag{5.1.3}$$

Of these terms c_2 can be conceived as containing the influence from the factors mentioned above, thus practically vanishing for distances larger than $2d$, where d is the maximum radius of a sample plot. It can then be assumed that c_1 is continuous and develops smoothly. Further c_1 (but not c_2) can be supposed to be independent of the size of the sample plot. It is therefore possible to rank different designs according to the variance per sample plot solely by considering the influence of c_1. Of course, it must then be presumed that the sample plots are equal in all the plans, and further that in all cases the distance between sample plots is large in comparison with d.

After comparing some types of design as to their efficiency in estimating the mean in the limiting case (§§ 5.2—5.4), we shall comment on the "small sample" problems (§ 5.5). In this connection, some data from experimental samplings of maps will be presented (§ 5.6).

As already mentioned, the choice of a sample design is depending not only on the precision of the estimates but also on the costs of sampling. When different designs are compared, it is, of course, especially important to consider those items of cost which depend on the design. In field sampling — as distinguished from the sampling of maps and the sampling of small surfaces in the laboratory — the cost of travel may be important. This cost is roughly proportional to the total distance covered when all sample points are visited. Some results concerning the air-line distances which must be covered when visiting the sample points will be presented in § 5.7. Although these investigations are not concerned with the stochastic processes sampled, they are intimately connected with the geometric questions that seem to emerge in almost all discussions of planar and spatial processes.

A unified treatment of the strategy of choosing a sample plan with due regard to all items of debit and credit will not be attempted. However, the final section (5.8) contains a short discussion of the efficiency of the various sample schemes. Here the previous results concerning precision and length of travel are utilized.

For the fundamental aspects of sampling the reader is referred to textbooks on the theory and method of sample surveys, such as Cochran (1953), Hansen, Hurwitz & Madow (1953), Yates (1953). Questions more directly connected with the theme of the present chapter can also be found in these books, especially in conjunction with the treatment of systematic sampling. As to other contributions references are found in the textbooks and in the author's earlier paper (1947). Further, a review of the literature of systematic sampling by Buckland (1951) should be especially mentioned. It is mainly confined to English and American papers. Contributions have also been made by Indian authors, see for example B. Ghosh (1949) and Das (1950). Among recent investigations Hájek (1959), Milne (1959), and Zubrzycki (1958) may be mentioned.

5.2. Size and shape of strata in stratified random sampling

The simplest type of random sampling is the *unrestricted random sampling*, also called *simple random sampling*. In the present case this type of sar .pling means that each sample point is selected with a uniform probability distribution over the whole of Q, independently of the selection of the othe: sample points. In the limiting case the variance per sample point is $c(0)$, where $c(v)$ is the cov.f. of the process $z(x)$. For the sake of simplicity it is here assumed that the process has no spectral mass at the origin. The corresponding component of the process would not have made any contribution to the variance (5.1.2) of any sampling scheme.

We then pass to the *stratified random sampling*. We assume that the region to be surveyed can be subdivided into a number of non-overlapping congruent *strata*. From each stratum k sample points are chosen at random. Let q stand for one particular stratum and let $\sigma_p^2(q)$ denote the variance per sample point, cf. (5.1.2). This quantity is uniquely determined, for the orientation of q does not affect the variance owing to the assumption of isotropy. We have

$$\sigma_p^2(q) = c(0) - D^2[z(q)] \qquad (5.2.1)$$

Here, in accordance with Cramér (1945, p. 180) D^2 denotes variance. Whereas $\sigma_p^2(q)$ is the variance within stratum, $D^2[z(q)]$ is the *variance between strata* (in an infinite region). By means of the formulas in § 2.5

$$D^2[z(q)] = \int_0^\infty c(v)\, b(v; q)\, dv \qquad (5.2.2)$$

where $b(v; q)$ is the frequency function of the distance between two points chosen at random and independently in q.

Under the assumptions made on $c(v)$, see § 5.1, q must be small and compact in shape, if $\sigma_p^2(q)$ shall attain a low value.

Let q and q_1 be regions of the same form with areas A and A_1 respectively. Then

$$D^2[z(q_1)] = \int_0^\infty c\left(v\sqrt{A_1/A}\right) b(v; q)\, dv \qquad (5.2.3)$$

From (3) we immediately conclude: If $A_1 \geqslant A$, then

$$\sigma_p^2(q_1) \geqslant \sigma_p^2(q)$$

Hence, if the shape of the stratum and the sampling intensity are given, the lowest sampling error is obtained if the stratum is made to contain one sample point only. As indicated in Ch. 6, however, we may for other reasons want at least two sample points per stratum.

It would be of interest to find mathematically precise formulations for the *optimum shape* of the stratum under various conditions. We shall however mainly confine ourselves to numerical illustrations showing the influence of the shape of the stratum on the variance.

By partial integration of (2) and inserting in (1)

$$\sigma_p^2(q) = -\int_0^\infty [1 - B(v; q)]\, dc(v) \tag{5.2.4}$$

Here B stands for the d.f. corresponding to the frequency function $b(v; q)$. According to the assumptions $-dc(v) > 0$. Therefore: If

$$B(v; q_1) \geqslant B(v; q_2)$$

for all v, then

$$\sigma_p^2(q_1) \leqslant \sigma_p^2(q_2)$$

The above inequality between the distribution functions is valid for example if q_1 is a circle and q_2 is a square of the same area as q_1; also if q_1 is a rectangle and q_2 is another rectangle of the same area but with longer diagonal. This is found by a numerical study of the distributions given in (2.5.18) and (2.5.20). If (2.5.22) and (2.5.23), too, are used, the following ranking of regions with equal area is obtained:

Circle
Regular hexagon
Square
Equilateral triangle

We shall then make use of the equality (2.5.19)

$$b(v; q) = \frac{2\pi v}{A} - \frac{2Pv^2}{A^2} + \dots$$

where A is the area of q and P is the perimeter. From this development some simple and obvious conclusions can be drawn.

Consider the class of plane figures of given area, for convenience here taken as unity. Further, fix a certain class of isotropic cov.f.'s, that may consist of all non-increasing functions, or may be a subclass that contains all completely monotone functions or all convex functions. In the class of functions there are members which are decreasing very rapidly towards 0, such as $\exp(-av)$ with a high value of a. For such a function only that part of $b(v; q)$ which is nearest to the origin has any influence on $\sigma_p^2(q)$. Thus, it can be concluded from (2.5.19) that of two regions with unit area the one with the shortest boundary has the lowest variance. If there is any region optimal for the whole class of cov.f.'s, it must therefore be the circle. However, it seldom happens

Table 5. Characteristics of seven regions of area 1.

Region (q)	Perimeter	$\int v\, b(v;\, q) dv$	$\int v^2\, b(v;\, q) dv$
Circle	3.545	0.5108	0.3183
Regular hexagon	3.722	0.5126	0.3208
Square	4.000	0.5214	0.3333
Equilateral triangle	4.559	0.5544	0.3849
Rectangle $\alpha=2$	4.243	0.5691	0.4167
Rectangle $\alpha=4$	5.000	0.7137	0.7083
Rectangle $\alpha=16$	8.500	1.3426	2.6771

that the region to be surveyed consists of a number of disjoint circles. It is more appropriate to deal with such strata that can form a network (mosaic), covering R_2 without overlapping. As is well known (Fejes Tóth 1953, p. 70), the regular hexagon is then optimal in the sense that it has the shortest boundary. We conjecture that it is optimal in relation to the whole class of non-increasing isotropic correlation functions.

As already indicated, this line will not be pursued further. It will be clear from examples that the difference in shape between figures such as the circle, the regular hexagon, and the square entails only unsignificant differences in variance. Although mathematically interesting, the problem of the optimal shape of a stratum, has therefore no immediate practical consequence.

If the correlation is rapidly decreasing, the area and perimeter of q suffice to characterize the performance of q as a stratum. At the other extreme, if the cov.f. is falling rather slowly (in relation to the size of the stratum), the lowest moments of $b(v;\, q)$ may be thought of as suitable characteristics. The first two moments and the perimeters of figures of the type already discussed are presented in table 5. As to the notation for rectangles, the shape is characterized by the ratio between the lengths of the sides. Throughout the chapter this ratio is denoted α. The reason for including the equilateral triangle, the square, and the regular hexagon in the table and elsewhere in this chapter, is that these three figures are the only regular polygons which can form a network over R_2 (see e.g. Sommerville 1929).

With one exception concordant rankings are obtained whichever characteristic is used as basis. However, it is evident that the rankings according to the moments and the perimeter may be quite different if non-convex regions are included.

We then proceed to numerical illustrations. We consider first the case of an exponential correlation. In table 6 is shown the variance per sample point in a number of cases when the cov.f. is $\exp(-v)$. It may be noticed that by aid of (3) also variances in some additional cases can be obtained from the table. Let σ_p^2 be the variance in the case of cov.f. $\exp(-v)$ and a stratum of area A. The variance remains the same when the cov.f. is replaced by $\exp(-hv)$ if at the same time the stratum is changed into a figure of similar shape and area A/h^2.

To indicate briefly how the values in table 6 and those of the subsequent table 7 were computed we take as example the case of circular strata and exponential correlation. The variance for a stratum of radius r is

$$1 - \psi_C(r)$$

where

$$\psi_C(r) = \int_0^2 \exp(-rv) f(v) \, dv \tag{5.2.5}$$

Here $f(v)$ is the frequency function of the distance between two random points in a circle of radius 1, see (2.5.20). For low values of r (5) was computed by means of a series development

$$\psi_C(r) = 1 - rm_1 + \frac{r^2}{2!} m_2 - \ldots \tag{5.2.6}$$

Here $m_1, m_2, \ldots$, are the moments of $f(v)$. Explicit expressions were obtained for the moments in all cases studied. Numerical integration was used for high values of r, and for the cov.f. used in table 7. The number of ordinates was 30—100 and the integration procedure was usually the "3/8th-rule". The major part of the computations was performed with the aid of the electronic computer Facit EDB at the Board of Computing Machinery in Stockholm. As in other cases, mentioned in subsequent sections, the programs were written by the author in the "alpha code" of Autocode Co.

It is immediately apparent from table 6 that the circle, the regular hexagon and the square give equivalent values of the variance. For a circular stratum

Table 6. Variance $\sigma_p^2(q)$ for different strata. Covariance function: $\exp(-v)$.

Area of stratum	Variance when the stratum is						
	Circle	Regular hexagon	Equilateral triangle	Rectangle (α is ratio between sides)			
				$\alpha = 1$ (square)	$\alpha = 2$	$\alpha = 4$	$\alpha = 16$
1/100	0.0495	0.0497	0.0536	0.0505	0.0549	0.0680	0.122
1/4	0.220	0.221	0.235	0.224	0.239	0.283	0.436
1	0.383	0.384	0.404	0.388	0.408	0.464	0.629
4	0.598	0.599	0.619	0.604	0.622	0.669	0.791
16	0.803	0.804	0.815	0.807	0.815	0.838	0.896
64	0.929	0.929	0.933	0.930	0.932	0.937	0.955

the variance is between 98.0 per cent and 100.0 per cent of the variance obtained when an equally large square stratum is used. The minimum value, 98.0 per cent, is attained when the area approaches zero. Hence, it equals the ratio between the mean values of the corresponding distributions of distance between random points, as is clear from (6). The same bounds, 98.0 per cent and 100.0 per cent, are valid for the whole class of completely monotonic covariance functions, cf. (2.4.10). The regular hexagon is seen to be intermediate between the circle and the square. It is further seen that the rectangle with the long side twice the length of the short side ($\alpha = 2$) has a variance of between 100.0 per cent and 109.2 per cent of that obtained for an equally large square. Similar computations for the rectangle with $\alpha = 1.5$, a case not included in table 6, gave the bounds 100.0 per cent and 103.2 per cent when comparisons are made with the square. Thus, the rectangles that are only moderately oblong are not much less efficient than the figures that are more compactly shaped. However, in some cases, the efficiency of a very elongated rectangle can be very poor in comparison with that of the square or the regular hexagon. Calculations show that a rectangle with $\alpha = 16$ can be less efficient than a square of six times larger area. As might be expected from table 5, an equilateral triangle is about as efficient as a rectangle that has the same area and where one side is twice the length of the other. It is further evident that the differences in variance between the figures are rather small when the correlation is rapidly falling in relation to the size of the strata. The bottom rows of table 6 show that in this case the stratified sampling is only slightly more efficient than the simple random sampling.

The function $\exp(-v)$ has a cusp at the origin, a trait characteristic of areal distributions, see 3.5 and 4.2. It seemed desirable to include also an isotropic correlation with zero derivative at the origin. The process with cov.f. $\exp(-v^2)$ is computationally easy. However, it was thought to be "too continuous" to be realistic. In fact it is deterministic along any straight line in R_2, see Karhunen (1952). Instead, the function $v\, K_1(v)$ was chosen. This function is suggested in Whittle (1954), see the remark following formula (2.4.9). It may be considered as intermediate between $\exp(-v)$ and $\exp(-v^2)$; in the vicinity of the origin $1 - v\, K_1(v)$ is of the order of magnitude $v^2 \log v$. Fig. 11 shows this function and the cov.f. $\exp(-v)$ of table 6.

Variances computed with the cov.f. $v\, K_1(v)$ are shown in table 7. The comparisons between different strata produce about the same results as those obtained in the case of exponential correlation. The maximum difference between variances of two different figures, however, is somewhat larger in the present case.

We conclude this section with a brief remark about *elliptic correlation* (see formulas 2.4.14—16). For the sake of convenience we assume that the process

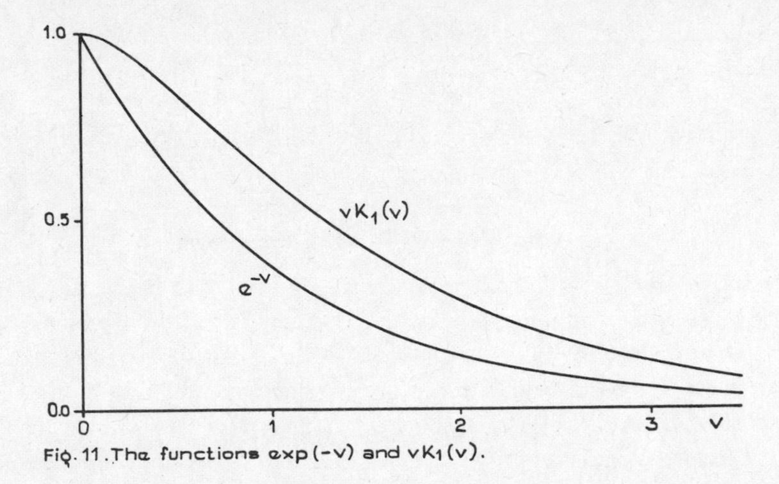

Fig. 11. The functions exp (-v) and vK1 (v).

$z(x, y)$ has strongest correlation along the x-axis. A simple compression of the xy-plane transforms z into an isotropic process z_1 defined by the equation

$$z_1(x, y) = z\ (bx, y)$$

where $b < 1$. Let now ξ be a random point in the region q. Then

$$z(\xi) - z(q) = z_1(\xi_1) - z_1(q_1)$$

where ξ_1 is a random point in q_1; q_1 is the mapping of q by the transformation $x_1 = bx,\ y_1 = y$. Thus, if Q is the region to be surveyed, the effectiveness of various strata in estimating $z(Q)$ can immediately be referred to the corresponding quantities in the isotropic case, viz. the case of estimating $z_1(Q_1)$, where Q_1 is the mapping of Q by the above transformation.

A simple example may suffice as an illustration. It was found that the square was nearly optimal in the isotropic case. Now if q_1 is a square in the x_1y_1-plane, its inverse image obtained by the above transformation is a parallelogram

Table 7. Variance $\sigma_p{}^2(q)$ for different strata. Covariance function: $vK_1(v)$.

Area of stratum	Variance when the stratum is						
	Circle	Regular hexagon	Equi-lateral triangle	Rectangle (α is ratio between sides)			
				$\alpha = 1$ (square)	$\alpha = 2$	$\alpha = 4$	$\alpha = 16$
1/4	0.0698	0.0701	0.0791	0.0720	0.0826	0.115	0.248
1	0.180	0.181	0.199	0.185	0.206	0.265	0.460
4	0.390	0.391	0.417	0.397	0.423	0.493	0.676
16	0.661	0.662	0.683	0.667	0.684	0.726	0.830

elongated in the x-direction. The simplest case is that of a square with the sides parallel to the axes. The corresponding "nearly optimum" stratum for the process with elliptic correlation is a rectangle with one side along the x-axis and the other along the y-axis, and the ratio $1/b$ between the lengths of these two sides.

5.3. Sampling by a latin square design

This section will briefly deal with a case of what has been called "deep stratification" (see Tepping, Hurwitz & Deming 1943, Patterson 1954), viz. sampling by means of the latin square design. It is supposed that the area to be surveyed consists of a number of congruent rectangular blocks. Each block is subdivided into n^2 "plots" of rectangular form and size $A \times B$. They are arranged in columns and rows with n plots in each row and n plots in each column. From each block n sample points are selected by the following two-stage procedure. First a cluster of n plots is chosen at random, with the restriction, however, that each column and each row is represented by one plot. In the second stage one sample point is selected at random from each one of the n plots chosen in the first stage.

The variance per sample point is

$$c(0) - \int_0^\infty c(v)\, a(v)\, dv \tag{5.3.1}$$

As before $c(v)$ signifies the cov. f., assumed to be isotropic. The "distance function" $a(v)$ is given by

$$a(v) = \frac{1}{n-1}\left[n\, a(v; nA, B) + n\, a(v; A, nB) - n\, a(v; nA, nB) - a(v; A, B)\right] \tag{5.3.2}$$

The symbol $a(v; s, s')$ is used for the frequency function of the distance between two random points in a rectangle with side-lengths s and s' (see 2.5.18). The right-hand side of (2) is derived by straightforward considerations of the probabilities involved in the sampling scheme.

By (2.5.19)

$$a(v) = \frac{2\pi v}{nAB} - \left(\frac{2v}{nAB}\right)^2 (A+B) + \dots \tag{5.3.3}$$

for small values of v. The first term is the same as that found in stratified sampling with the same sampling intensity and one point per stratum. If the sampling ratio is one point per unit area, (3) is simply

$$2\pi v - 4(A+B)\, v^2 + \dots$$

Table 8. The $n \times n$ latin square. Plots with sides A and B. One sample point per unit area. Values of the moment $\int v\ a(v)\ dv$.

n	$A/B=1$	$A/B=2$	$A/B=4$	$A/B=16$
2	0.433	0.458	0.543	0.961
3	0.420	0.433	0.486	0.798
4	0.426	0.432	0.463	0.705
6	0.456	0.452	0.455	0.605
8	0.492	0.481	0.466	0.553
12	0.564	0.546	0.506	0.508
16	0.631	0.607	0.550	0.494

If $c(v)$ is rapidly falling it would hence be advantageous to use a small value of $(A+B)$, which can be obtained by choosing a high order (n) of the latin square, and by making the plots square in shape. However, it will be clear from the results presented in table 9, that these considerations apply only in cases where the cor.f. is decreasing so fast that the first term in (3) is of dominating importance when compared with the following terms. As indicated in § 5.2, in this case no sampling scheme can give any substantial advantage over the simple random sampling.

The performance in the case of a slowly decreasing correlation is of higher interest. Then (1) is dependent upon the whole course of the distance function. However, in contrast to the corresponding function in stratified random sampling (2) has a complicated course with change of sign. It was thought, however, that the first moment

$$\int_0^\infty v\ a(v)\ dv$$

would give some indication of the effectiveness of the various designs, since the variance (1) is proportional to this quantity in the limiting case when $h \to 0$ in the cor.f. $r(v) = \exp(-hv)$. The first moment has therefore been tabulated in a few cases (table 8). It should be noticed that the corresponding moment in stratified sampling with square stratum is 0.5214 (table 5). It is seen that this quantity is also obtained as a limiting value when the order of the latin square tends to infinity and when the shape of the plot is adjusted so that $\int v\ a(v)\ dv$ is made as small as possible $\left(\text{i.e. a plot approximately } 1 \times \frac{1}{n}\right)$. It is further seen from table 8 that in some cases a substantial reduction (in comparison with the best stratified design) is attained. See e.g. the plans of order 2×2 to 4×4 with square or nearly square plots.

Table 9 contains some variances for latin square designs with square plots. The cov.f.'s are of the type used in the computations of the preceding section. Two cases of stratified sampling are included for reference as well as a number of systematic sample schemes to be discussed in the next section. For the case of square plots it may evidently be concluded that latin squares of low order, i.e. $\leqslant 6\times 6$, give a smaller variance than the stratified samples. However, further investigations are required if more definite conclusions are to be reached.

5.4. Some cases of systematic sampling

In systematic sampling the cluster of sampling units forms some regular geometric pattern. We shall here only discuss some special cases of sampling by a lattice of points in a parallelogram arrangement. Thus, it is understood that the sample points $(x_{\mu\nu}, y_{\mu\nu})$ are given by a relation of the type

$$\begin{aligned} x_{\mu\nu} &= a(\xi+\mu) + b(\eta+\nu)\cos\varphi \\ y_{\mu\nu} &= b(\eta+\nu)\sin\varphi \end{aligned} \tag{5.4.1}$$

Here a, b, and φ are fixed constants. The sampling procedure is as follows: a point (ξ, η) is first chosen in the unit square. As sample points are then used all those lattice points which are given by (1) and belong to the region under survey. The scheme (1) is restricted in so far as it has one side of the basic parallelogram parallel to the x-axis. However, this restriction is of no consequence when isotropic processes are dealt with.

Since we shall neglect the border effect, as mentioned in § 5.1, the variances will be the same if the set of sample points is chosen without any random procedure, e.g. by fixing in advance $\xi=\eta=0$, or by centering the sample points with reference to the boundary of the region (Yates 1948, Milne 1959).

We first note that the number of sample points per unit area is

$$\frac{1}{ab\sin\varphi}$$

For the sake of convenience the sampling ratio giving 1 point per unit area will be applied in the sequel. Thus $ab\sin\varphi=1$. We notice the following special cases:

The triangular (or hexagonal) network, $a=b=\sqrt[4]{4/3}$, $\varphi=\pi/3$. The six points closest to any given point in the network are the corners of a regular hexagon with the given point as center.

The square network, $a=b=1$, $\varphi=\pi/2$.

The modified square network, $a=1$, $b=\sqrt{5/4}$, $\operatorname{tg}\varphi=2$. This system is obtained from the square network by moving every second of the lines parallel to the x-axis one half step to the left or to the right.

The general rectangular case, $ab=1$, $\varphi=\pi/2$. In dealing with this case the ratio a/b will be denoted α.

We shall now study the performance of systematic schemes when the same type of processes as those treated in the earlier sections of this chapter are sampled. First the formulas which have been used in the calculations will be presented.

We then start with the notationally simplest case, the square network. To avoid complications at the borders, it is assumed that the region under survey, Q, is a rectangle with the sides parallel to the axes and the size $m \times n$, where m and n are integers. The variance per sample point is then

$$mnE\left\{\left[\frac{1}{mn}\sum_{1}^{m}\sum_{1}^{n} z(\mu+\xi, \nu+\eta) - z(Q)\right]^2\right\} =$$

$$= \frac{1}{mn} D^2[\Sigma\Sigma z(\mu, \nu)] - mn\, D^2[z(Q)] \qquad (5.4.2)$$

Expressing (2) in the cov.f. it is convenient to use for this function a notation valid also in the non-isotropic case. Rearranging the terms of (2) we find

$$\sum_{-m}^{m}\sum_{-n}^{n}{}'(1-|\mu|/m)(1-|\nu|/n)\,c(\mu,\nu) -$$

$$-\int_{-m}^{m}\int_{-n}^{n}(1-|x|/m)(1-|y|/n)\,c(x,y)\,dx\,dy$$

When m and n tend to infinity, this expression approaches

$$\sigma_p^2 = \sum_{-\infty}^{\infty}\sum_{-\infty}^{\infty} c(\mu,\nu) - \int_{-\infty}^{\infty}\int_{-\infty}^{\infty} c(x,y)\,dx\,dy \qquad (5.4.3)$$

Assume now for simplicity that $c(0, 0) = 1$. If further the spectral distribution is absolutely continuous, (3) can be expressed in the spectral density. Denoting this density by $f(x, y)$, Poisson's formula gives (Bochner 1932, p. 203)

$$\sigma_p^2 = 4\pi^2\left[\sum_{-\infty}^{\infty}\sum_{-\infty}^{\infty} f(2\pi\mu, 2\pi\nu) - f(0,0)\right] \qquad (5.4.4)$$

It should be observed that even in the limiting case the problem is not that of estimating the a priori mean $E[z(x, y)]$ of the process. In (4) it is exactly the subtraction of $f(0, 0)$ that constitutes the distinction.

When the cov.f. is falling rapidly, as $\exp(-hv)$ with a high value of h, it is easy to evaluate the variance from (3) since only a few terms of the double sum have to be computed and the value of the integral is directly obtained; it equals $2\pi/h^2$ in this case.

With this exception, (4) has been used in the computations. The relevant spectral densities have been given in § 2.4. The spectral density corresponding to the cov.f. $\exp(-hv)$ is

$$\frac{h}{2\pi}(h^2+x^2+y^2)^{-3/2} \tag{5.4.5}$$

whereas the density corresponding to $bv\,K_1(bv)$ is

$$\frac{b^2}{\pi}(b^2+x^2+y^2)^{-2} \tag{5.4.6}$$

The computations with f given by (5) were carried out on the computer Facit EDB already mentioned. The calculations consisted of a straightforward summation for all lattice points within a certain radius (of the order of magnitude 20—60) from the origin, and an approximation by integration for the points outside this radius. The accuracy was checked by varying the radius.

The density (6) is rational. In this case, the summation was first carried out row-wise by means of formulas for the decomposition into partial fractions of the hyperbolic functions. The ensuing summation by columns was reduced to adding a small number of relevant terms.

Returning to the general case (1) we introduce the process

$$z_1(x, y) = z(ax + by\cos\varphi,\ by\sin\varphi) \tag{5.4.7}$$

Sampling z by the scheme (1) is equivalent to sampling z_1 by a square network. Thus, (3) and (4) apply if the cov.f. and the spectral density (respectively) of z_1 are inserted in the formulas. These functions are given by the following equations

$$c_1(x, y) = c(ax + by\cos\varphi,\ by\sin\varphi) \tag{5.4.8}$$

$$f_1(x, y) = (1/ab\sin\varphi)\ f\left(\frac{x}{a}, \frac{ay - bx\cos\varphi}{ab\sin\varphi}\right) \tag{5.4.9}$$

where c and f are the corresponding functions for z.

The program written for the computer in the case (5) was made to include summation of any expression of type (9). The methods indicated above work also when (9) is applied to (6).

In this way the values in the lower half of table 9 were obtained. It is seen that the triangular network gives the highest precision of all schemes included in the table—for all values of h and b in (5) and (6). However, the modified square network and the square network give variances which are but slightly higher than those obtained in the triangular system. It is also seen in some cases that the gain over all the random schemes is considerable.

Table 9. Variance for different sample designs with one sample point per unit of area. The variance in unrestricted random sampling = 100.

Sample design	Variance in per cent of unrestricted random sampling when the covariance function is								
	$\exp(-hv)$ with $h =$					$bvK_1(bv)$ with $b =$			
	0.1	0.5	1	2	4	0.5	1	2	4
Stratified									
(square strata)									
1 point per stratum	5.05	22.4	38.8	60.4	80.7	7.20	18.5	39.7	66.7
2 points per stratum	7.05	29.8	49.2	71.2	88.0	11.8	27.8	53.2	78.3
Latin square									
(square plots)									
2 × 2	4.25	19.6	35.2	57.1	79.0	5.24	14.8	35.1	63.8
4 × 4	4.21	19.8	35.9	58.0	79.3	5.12	15.2	36.4	64.4
6 × 6	4.52	21.1	38.0	60.1	80.3	5.85	17.2	39.4	66.4
16 × 16	6.26	28.2	47.3	68.0	84.1	10.5	26.9	51.0	73.5
Systematic									
Triangular	2.25	11.2	22.1	41.7	69.4	1.16	4.50	16.3	46.1
Modified square	2.26	11.3	22.2	41.9	69.5	1.17	4.55	16.4	46.3
Square	2.29	11.4	22.4	42.2	69.7	1.19	4.68	16.8	46.8
Rectangular									
$\alpha = 2$	2.90	14.4	27.8	50.0	74.9	2.08	7.86	25.5	57.0
$\alpha = 4$	5.70	27.7	51.3	79.8	92.3	6.85	24.1	62.4	90.2
$\alpha = 8$	14.3	67.6	115	144	125	25.5	79.6	152	149
$\alpha = 16$	39.2	174	258	252	177	93.3	238	319	239
$\alpha = 64$	306	1015	978	647	369				

Meanwhile, it is seen by an argument used in discussing the optimum shape of a stratum (5.2) that the only system which can be optimum for a wide class of non-increasing isotropic correlations is the triangular network. For when the cor. f. is rapidly decreasing, the lowest variance is obtained for a system with the highest lower bound for the distance between two different sample points. The triangular system has this property (for a fixed sampling ratio), see Fejes (1940).

It is further seen from table 9, that even a moderate deviation from the square pattern such as a rectangular system with $\alpha = 2$ (i.e. the system $a = \sqrt{2}$, $b = \sqrt{1/2}$) can give a substantial increase in the sampling error. In extreme cases such as those with $\alpha = 16$ and $\alpha = 64$ the sampling errors may be much higher than those occurring in unrestricted random sampling. These extreme cases are of the type known as "line-plot survey" in forestry.

5.5. Some remarks about the case of small samples

When the total number of sample units is small, certain "border effects" appear. Effects of this kind may be encountered even in a large-scale survey when estimates for subregions of the total region under survey are required. Such a subregion may be of a rather complicated nature, consisting for example of a large number of scattered disjoint domains.

We shall here deal with the same sample schemes as in the preceding sections. Let us consider for example a network of congruent strata. Assume that this network is placed at random over the plane without any consideration of the pattern which the network forms over the particular subregion. The corresponding assumptions are made in the case of designs of the latin square or the systematic type. Problems such as the determination of the optimal arrangement of four sample points in a square region, etc., will not be discussed at all. An attempt to find general results of practical interest would evidently lead into an enumeration of a very large number of particular cases.

The case chosen for discussion here corresponds fairly well to the situation in forest surveys. A rigid frame of sampling units is often located without reference to the boundaries of the region surveyed. It seems realistic then to think of the frame of sampling units as randomly located over the area.

As to the general problems arising when estimates for a subpopulation are required from a sample which has been chosen without specific regard to the particular subpopulation, a discussion is found in Durbin (1958). In this context it may be mentioned that the following remarks concern other cases of plane sampling, too, such as "line sampling", "line-plot sampling", etc.

We now consider, as before, a stationary isotropic process $z(x)$ with $x \varepsilon R_2$. Some sample scheme of the types treated in the earlier sections of this chapter is assumed. We suppose that the scheme consists of a procedure for locating sample points at a given sampling intensity over a very large region Q_0, that contains the small region Q for which estimates are wanted. Instead of assuming that a frame of e.g. basic strata is placed at random over Q, it will be convenient to assume that the frame is fixed in advance and that Q is placed at random in Q_0. This is explained by the following procedure. A point P and a direction PB are fixed in the given figure Q. For the location of Q in Q_0 it is then assumed that P is placed in a point which is chosen at random with uniform probability over Q_0, and that the angle between PB and a fixed direction in Q_0 is selected at random with uniform probability over the interval $(0, 2\pi)$, cf. the kinematic measure in plane integral geometry (Santaló 1953, p. 21).

The inclusion of Q_0 in the considerations is, of course, merely a device intended to simplify the discussion. As to the assumptions on $z(x)$, it seems natural to suppose that the mechanism producing realizations of the process does not contain any long-wave components, i.e. long with respect to the size of Q. Thus the overall behaviour of $z(x)$ in Q would not be very much dependent upon the location of Q in Q_0.

We denote the areas of Q and Q_0 by A and A_0, respectively. We then introduce

$$e(x) = \begin{cases} 1 & x \varepsilon Q \\ 0 & \text{otherwise} \end{cases}$$

and define

$$z_1(x) = z(x)\, e(x)$$

Assuming that Q_0 is so large that its border effects can be neglected, we can regard $e(x)$ and $z_1(x)$ as stationary inside Q_0, where further $e(x)$ is independent of $z(x)$. Neglecting terms which are small in comparison with A/A_0, the following moment formulas are obtained for $x, y \varepsilon Q_0$

$$\left.\begin{aligned} &E[e(x)] = A/A_0 \\ &\text{Cov.}\ [e(x),\ e(y)] = \frac{A}{A_0} \cdot \frac{A f(v)}{2\pi v} \\ &E[z_1(x)] = mA/A_0 \\ &\text{Cov.}\ [z_1(x),\ z_1(y)] = \frac{A}{A_0} \cdot \frac{A f(v)}{2\pi v} [c(v) + m^2] \end{aligned}\right\} \qquad (5.5.1)$$

Here v denotes the distance $|x—y|$, whereas m and $c(v)$ are the mean value and the cov.f., resp., of the process $z(x)$. Moreover, $f(v)$ is the frequency function of the distance between random points in Q. The cov.f. for the process $e(x)$ is obtained by considering the conditional probability of the event $e(y) = 1$, given $e(x) = 1$.

Next, consider the estimation of the integral

$$Z(Q) = \int_{Q_0} z_1(x)\, dx$$

by means of the sum

$$Z = \sum z_1(x_\nu)$$

extended over all sample points in Q_0. For the sake of simplicity, it is here supposed that there is on the average one sample point per unit of area, thus A_0 sample points in Q_0. The number of sample points in Q is not fixed, it is a random variable depending on the size and shape of Q and the sampling plan used.

It is first noted that $E[Z - Z(Q)] = 0$, i.e. the estimate is unbiased. For the variance

$$E\{[Z - Z(Q)]^2\}$$

the formulas of the preceding sections can be applied. After multiplication by A_0, the cov.f. of z_1, given in (1), can be inserted in the previous formulas.

It would carry too far to discuss all sample plans. We shall therefore mainly deal with the notationally simplest cases, i.e. the unrestricted and the stratified random sampling.

For the first type of sampling we find

$$E\{[Z - Z(Q)]^2\} = A[m^2 + c(0)] \tag{5.5.2}$$

If the stratum is denoted by q, as in 5.2, the variance in the case of stratified sampling is (cf. 5.2.1 and 5.2.2)

$$A[m^2 + c(0)] - A\int_0^\infty [m^2 + c(v)] \frac{A f(v)}{2\pi v} b(v; q)\, dv \tag{5.5.3}$$

Assuming Q to be compact in shape and large in comparison with the stratum, q, we may use the expansion (2.5.19) for $f(v)$. Thus the following approximation is obtained where P denotes the perimeter of Q

$$E\{[Z - Z(Q)]^2\} \approx A\{c(0) - \int_0^\infty c(v)\, b(v; q)\, dv +$$

$$+ \frac{P}{\pi A}\int_0^\infty [m^2 + c(v)]\, v\; b(v; q)\, dv\} \tag{5.5.4}$$

The conclusion regarding the influence of the shape of Q on the variance is obvious. It is also clearly seen that the recommendations concerning the choice of q must chiefly be the same in the present case as those made in the limiting case (see 5.2).

The special case $m = 1$, $c(v) \equiv 0$, is equivalent to the estimation of the area of Q. In unrestricted random sampling the estimate has a Poisson distribution with the mean value A. In the stratified case the variance is

$$A\left[1 - A\int_0^\infty \frac{f(v)}{2\pi v} b(v; q)\, dv\right] \tag{5.5.5}$$

When the conditions underlying (4) are valid, we get the approximation

$$E[(Z - A)^2] = m_1 P/\pi \tag{5.5.6}$$

where m_1 is the mean distance between random points in q.

Formula (5) constitutes a rather simple expression for the way in which the sample points are distributed over Q_0 in stratified sampling. It is of some interest to compare this formula with the corresponding one in systematic sampling. When the sample points are given by the general formula (5.4.1) but with 1 sample point per unit area (i.e. $ab \sin \varphi = 1$) the variance in the estimate of A is

$$A^2\left\{\sum_{\mu=-\infty}^{\infty}\sum_{\nu=-\infty}^{\infty} \frac{f(w)}{2\pi w} - 1\right\} \tag{5.5.7}$$

where

$$w^2 = a^2\mu^2 + b^2\nu^2 + 2\,ab\mu\nu\cos\varphi$$

For $\mu = \nu = 0$, $f(w)/2\pi w$ should be replaced by its limit value $1/A$, cf. (2.5.19).

Some simple examples of variances computed from (5) and (7) are shown in table 10. A square stratum is assumed in the case (5), whereas a square network is chosen in (7). The variance in unrestricted random sampling, too, is included. It equals the true value of the area to be estimated.

Consider now the estimation of the average

$$z(Q) = \int_{Q_0} z_1(x)\,dx \Big/ \int_{Q_0} e(x)\,dx$$

As estimate of $z(Q)$ the following ratio is used

$$\bar{z} = \Sigma z_1(x_\nu) \,/\, \Sigma e(x_\nu)$$

with summation over all sample points in Q_0.

We consider first the unrestricted random sampling. If the number of sample points in Q is known, we have the conditional expectation

$$E\{[\bar{z} - z(Q)]^2\} = [c(0) - \int_0^\infty c(v)\,f(v)\,dv]\,/\,n \tag{5.5.8}$$

The number of sample points, n, has a Poisson distribution. In the truncated distribution where $n = 0$ is excluded, the expectation of $1/n$ can be approximated as

$$\frac{1}{A - 1 - 1/A} \tag{5.5.9}$$

The expected value of n is A, since we have assumed on the average 1 sample point per unit area. The exact value of $E(1/n)$ in the truncated distribution

Table 10. Estimating the area of a randomly located figure. Variances for three types of sampling. (Sampling ratio: 1 sample point per unit area.)

Figure	Variance in		
	Unrestricted random sample	Stratified random sample 1 point per square of area 1	Systematic sample Square network
Square with side $1/\sqrt{2}$	0.500	0.354	0.250
Square with side 1	1.000	0.557	0.180
Square with side $\sqrt{6}$	6.000	1.518	1.058
Rectangle $\sqrt{6} \times 1/\sqrt{6}$	1.000	0.695	0.564
Regular hexagon with side 1	2.598	1.957	0.619

is found from tables of the exponential integral, $\overline{Ei}(x)$, since (Jahnke & Emde 1945, cf. Grab & Savage 1954)

$$\sum_{j=1}^{\infty} \frac{x^j}{j \cdot j!} = \overline{Ei}(x) - C - \log x$$

Before treating the remaining designs we must introduce some restrictions. In spite of this the results will be approximations only. This is, of course, the common situation in dealing with ratio estimates. Assume that the sums

$$\Sigma e(x_\nu), \qquad \Sigma z_1(x_\nu)$$

show deviations from their expectations which are small in comparison with the expectations. The expected value of the first sum is A. Let the expectation of the second sum be denoted B. According to the assumptions made

$$\bar{z} - z(Q) \approx (1/A)\left\{\Sigma z_1(x_\nu) - B - \frac{B}{A}[\Sigma e(x_\nu) - A]\right\}$$

We introduce a new process by the equation

$$u(x) = z_1(x) - \frac{B}{A} e(x) \qquad (5.5.10)$$

Clearly $u(x)$ is stationary and isotropic. The following approximate expression is obtained for the variance of $\bar{z}$

$$E\{[\bar{z} - z(Q)]^2\} \approx (1/A^2) \ E(\{\Sigma u(x_\nu) - E[\Sigma u(x_\nu)]\}^2)$$

We can now use for example (3) and (4) if m and $c(v)$ are replaced by the corresponding characteristics of $u(x)$.

A similar device can be used in the more general case of estimating a ratio

$$\int_Q z_1(x)\,dx \Big/ \int_Q z_2(x)\,dx$$

where z_1 and z_2 are arbitrary, see Matérn (1947, pp. 79 ff.).

5.6. Empirical examples

This section will comprise two empirical examples, the first of which refers to the "small sample" case.

This example concerns four subregions cut out from a map on the scale 1 : 200,000 (Stockholms och Uppsala län samt Stockholms Överståthållarskap, Generalstabens litografiska anstalt, 1947). Administrative lines of demarcation form the boundaries of the regions. They are shown in fig. 12.

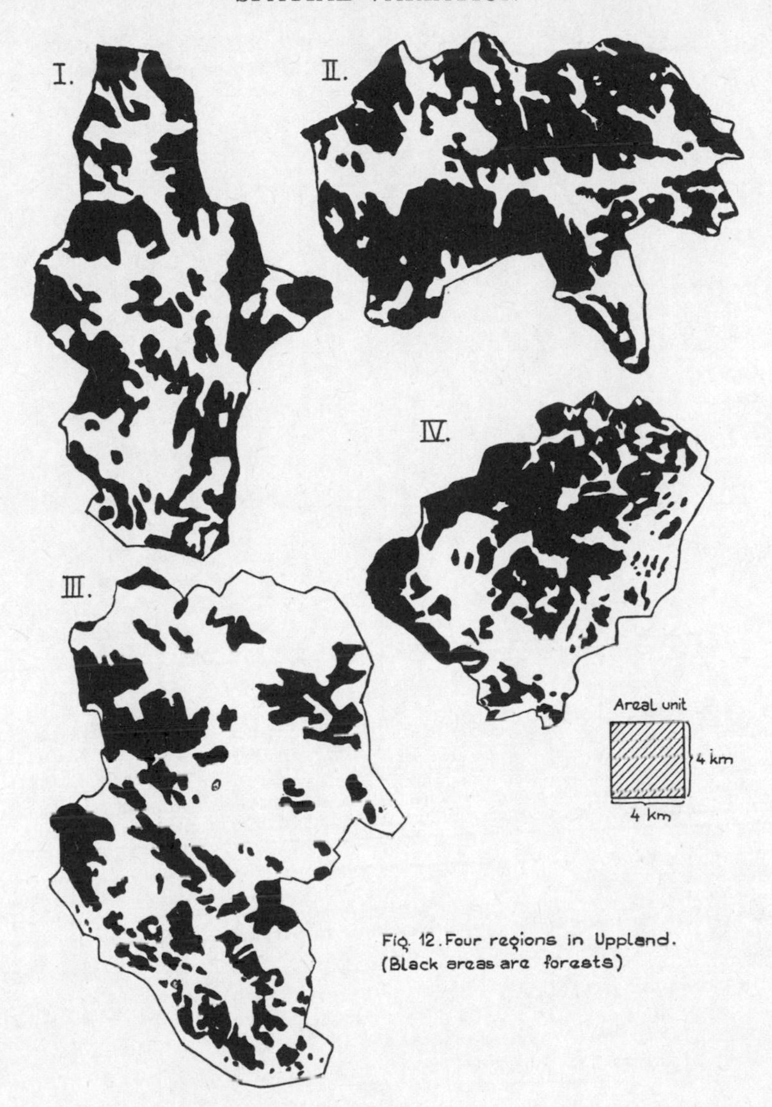

Fig. 12. Four regions in Uppland. (Black areas are forests)

Sampling experiments were performed with a sampling ratio giving on the average one point per unit area. The unit area was 4 sq. cm on the map (16 sq. km in the field). In such units the areas of the four regions are

I: 13.8 II: 15.6 III: 20.3 IV: 11.6

The results of the experiments and information about the sampling designs used are found in table 11. The values in the table are empirical variances, except those in the last column (G: unrestricted random sampling), which are theoretical expressions, belonging to the Poisson distribution or calculated from

Table 11. Sampling from a map of the province of Uppland. Variances estimated from experiments (A—F) or computed theoretically (G).

Estimate of	Sampling scheme						
	A	B	C	D	E	F	G
Total area							
Region I	1.32	1.19	2.03	2.29	3.08	1.86	13.8
II	1.59	1.19	7.17	2.99	5.14	3.00	15.6
III	1.54	1.28	3.22	2.46	5.09	2.28	20.3
IV	0.75	1.00	1.58	2.79	2.21	1.94	11.6
Average	1.30	1.16	3.50	2.63	3.88	2.27	15.3
Total forest area							
Region I	3.3	3.2	3.0	3.4	4.9	6.0	7.4
II	3.6	2.2	5.1	4.8	8.1	4.4	10.3
III	2.3	2.1	4.5	3.6	3.1	3.3	5.8
IV	1.9	2.9	4.0	4.5	4.3	2.0	6.3
Average	2.8	2.6	4.1	4.1	5.1	3.9	7.5
Total water area							
Region I	0.3	0.5	0.2	0.2	0.2	0.2	0.4
II	0.9	1.1	1.1	0.5	0.8	0.8	1.0
III	1.9	1.7	2.7	3.8	3.0	2.0	2.5
IV	1.5	0.8	2.4	1.7	2.1	0.8	2.7
Average	1.2	1.0	1.6	1.6	1.5	1.0	1.6
Forest area percentage							
Region I	113	128	199	119	162	273	195
II	116	80	110	78	169	129	154
III	61	54	86	81	73	93	106
IV	133	172	186	250	346	152	236
Average	106	108	145	132	188	161	173
Water area percentage							
Region I	17	27	12	10	10	11	22
II	35	43	35	22	38	30	40
III	42	43	63	72	65	45	56
IV	89	67	171	118	132	62	172
Average	46	45	70	56	61	37	72

A—C: Systematic samples, A) Triangular, B) Square, C) Rectangular $2 \times {}^1/_2$
D—E: Stratified samples, D) 1 point per unit square, E) Four points per 2×2 square
F: Latin square design, 4×4 square plots of size ${}^1/_2 \times {}^1/_2$
G: Unrestricted random sample

(5.5.8) and (5.5.9). For each of the remaining six sample plans, 25 independent experiments were carried out in every region. A basic frame of strata or of systematically located sample points was then oriented at random over the region, as described in § 5.5. The following observations were made in each experiment: the total number of sample points, the number of points in areas marked as forest land, and the number of points in areas marked as water. From these numbers also two percentages were computed: number of points in forest area and number of points in water area in per cent of the total number of points in the region.

It should be observed that the variances are based on 24 degrees of freedom only, for every particular region. Still, it might be of certain interest to compare for example the approximation (5.5.6) with the empirical variances found in the stratified sampling for comparatively small strata (scheme D with stratum = unit square). The constant m_1 in (5.5.6) equals 0.5214 (table 5). The perimeter values of the four regions were approximately

I: 18.6 II: 20.6 III: 22.2 IV: 15.3

Hence the following approximate variances for the estimated areas are obtained from (5.5.6)

3.0 3.4 3.7 2.5 (average 3.2)

whereas the corresponding empirical values of table 11 are

2.3 3.0 2.5 2.8 (average 2.6)

The agreement is fairly good. It is easily seen that (5.5.6) always must overestimate the variance.

In the second example a square grid of 50 × 50 points was located on the map (scale 1 : 50,000) which furnished three of the correlograms of table 2 (§ 4.2). The distance between successive points in the grid was in either direction 1 cm (0.5 km). For each of the 2,500 points it was noted whether situated on land or on water. Six different designs for selecting 25 points from the population of 2,500 points were applied. The designs are listed in table 12. In this case no experimental sampling was carried out. Each variance in the table expresses the variation among all possible outcomes of the particular sampling plan. It should be noted that no border effects are operating in this case. To take the third design as an example, the population of 50 × 50 points was divided into five exactly equal strata of 10 × 50 points.

Table 12. Sampling from the map "Stockholm SO" (1 : 50,000). Seven designs with the same sampling intensity. Variances in per cent of that in unrestricted random sampling.

Sampling scheme	Degrees of freedom for the estimate of variance	Variance
Unrestricted random sampling	2,499	100
Stratified random sampling:		
1 point per 10 × 10 square	2,475	68.6
5 points per each 10 × 50 rectangle	2,495	85.2
5 points per each 50 × 10 rectangle	2,495	92.2
Latin square design, 5 points per plot. 5 plots selected from 25 plots of 10 × 10 points	2,491	77.4
Systematic sampling. Square pattern (10 × 10)	99	64.9

The values in table 12 are variances expressed in per cent of the variance in unrestricted random sampling.

It can be shown that the variances in table 12 are in good agreement with those expected from the correlograms of table 2. This however should be almost self-evident, since the basic data are identical for table 12 and the correlograms.

5.7. Average travel distance between sample points

This section will present some results concerning the length of a route covering all sample points of one or the other sample scheme applied to a plane region. Only the limiting case when border effects can be neglected will be discussed. Further, by distance is meant air-line distance, as already mentioned in § 5.1.

The treatment of the systematic sampling plans presents no problems. It seems to be immediately clear in each case which path is the shortest one, see further the cases entered in table 13.

However, a random element present in the sampling introduces obstacles which appear almost unsurmountable, at least when the exact expectation of the shortest route is wanted. However, it would seldom be economically feasible to determine the permutation of the sample points that gives the exact minimum length. It would probably be of greater value to develop an easily applicable method of constructing a path which is (for practical purposes) sufficiently close to the minimum length.

This course will be followed here. Hence, a rigid rule of determining a path connecting the sample points will be assumed for each particular design, and the corresponding average length will be computed. However, it will usually be possible to shorten the path by occasional deviations from the rule, deviations which are easily conceived once the sample points are selected. Therefore the values presented here for the random schemes will generally be higher than those which are attained in practice.

We consider first the stratified sample with rectangular strata and one sample point per stratum. We suppose that the strata are arranged in rows, and that the strata forming a row have contact along their long sides, which are assumed to be parallel to the y-axis. The sample points in every row can be arranged in a sequence according to ascending x-coordinates. The path chosen here covers the points of every second row in the above order, and the points of the intermediate rows in the opposite order.

Let the sides of the rectangle have lengths a and b with $a \geqslant b$. The average distance from one sample point to the next one is

$$2\,M(a,\,2\,b) - M(a,\,b) \tag{5.7.1}$$

where $M(a, b)$ is the expected distance between two points chosen at random and independently in a rectangle with sides of the lengths a and b. The expression (1) is derived by elementary calculus of probability. For easy reference we give here the function $M(a, b)$ (see Ghosh 1951)

$$M(a, b) = \sqrt{ab}\,[m(a/b) + m(b/a)] \tag{5.7.2}$$

wherein

$$m(t) = \frac{1}{30\sqrt{t}}\,[2t^3 + (3 - 2t^2)\sqrt{1+t^2} + 5t^2 \text{ Arsinh } (1/t)]$$

Values computed from (2) have been given in table 5 of § 5.2. The formula was also employed at the calculation of table 8 in § 5.3.

Similar serpentine routes have been presupposed for all random designs entered in table 13.

Only one case of stratified sampling with more than one point per stratum is considered here, viz. the selection of two points from each square stratum of size $\sqrt{2} \times \sqrt{2}$. Here, too, the strata are supposed to be arranged in rows, and the points in a row are covered in order of ascending or descending x-coordinates. This means that the average entered in table 13 is computed by the formula

$$\frac{1}{2}\sqrt{2}\left[M(1, 1) + \int_0^2 g(x)\,dx \int_0^1 \sqrt{x^2+y^2}\,h(y)\,dy\right]$$

with

$$g(x) = \begin{cases} 4x(1-x) + 2x^3/3 & 0 < x < 1 \\ 2(2-x)^3/3 & 1 < x < 2 \end{cases}$$

$$h(y) = 2(1-y)$$

Here g and h are frequency functions with the following meaning. Two unit squares have a common side along the y-axis. In each one two random points are selected. The point in the ith square that is closest to the y-axis is denoted (x_i, y_i). Then g and h are the frequency functions of the distances $|x_1 - x_2|$ and $|y_1 - y_2|$, respectively.

The two latin square designs of table 13 refer to the case of square plots. It is assumed that the latin squares are arranged in rows, and that the points in a row are covered as described above. The averages are easily obtained as linear expressions in values of $M(a, b)$.

We now pass to the unrestricted random sampling. This seems to be the most intriguing case of those considered here. We assume that there is on the average one sample point per unit area. Since the border effects are neglected, the system of sampling points can be regarded as a realization of a plane Poisson process with intensity 1.

Table 13. Average length per sample point of a path passing through all sample points in a plane region. Sample ratio: 1 point per unit area. Limits when the area of the region tends to infinity.

Sampling scheme	Average length
Systematic	
Triangular network	1.0746
Square network	1.0000
Rectangular network $\sqrt{2} \times \sqrt{1/2}$	0.7071
Rectangular network $2 \times 1/2$	0.5000
Stratified, 1 point per stratum	
Stratum square 1×1 ($\alpha = 1$)	1.0881
Stratum rectangle $\sqrt{2} \times \sqrt{1/2}$ ($\alpha = 2$)	0.9057
Stratum rectangle $\sqrt{3} \times \sqrt{1/3}$ ($\alpha = 3$)	0.8792
Stratum rectangle $2 \times 1/2$ ($\alpha = 4$)	0.8958
Stratified, 2 points per stratum	
Stratum square $\sqrt{2} \times \sqrt{2}$	0.9261
Latin square, square plots	
2 points in a 2×2 square (plot $\sqrt{1/2} \times \sqrt{1/2}$)	0.9741
4 points in a 4×4 square (plot $1/2 \times 1/2$)	1.0192
Unrestricted random	0.9212

Before proceeding to the construction of the rules for covering all the sample points, we give some indications about certain investigations related to the present problem.

Several authors have dealt with different aspects of the shortest path through a number of points in a plane region. We shall here only be concerned with the conclusion which can be drawn for the present problem from the published results. We denote the average length of a segment of the shortest path (i.e. a straight line joining two sample points) by L. From Few's investigation (1955) can be concluded that $L \leqslant \sqrt{2}$ for any location of the sample points, as soon as the overall sampling intensity is 1. According to a conjecture by Verblunsky (1951) the upper bound is 1.075, i.e. equal to the distance between neighbouring points in a hexagonal lattice. For the expected value of L in simple random sampling the following inequality is found from Marks (1948) and M. N. Ghosh (1949)

$$0.50 < E(L) < 1.27 \tag{5.7.3}$$

Ghosh, who established the upper bound, partly based his result on artificial sampling. This bound is outside the range of possible L-values if Verblunsky's conjecture is correct.

Now it is possible by simple means to replace (3) with a somewhat sharpened inequality, viz.

$$0.625 < E(L) < 0.922 \tag{5.7.4}$$

The lower bound is immediately obtained by observing that the total length of the two sections of the shortest path through one particular sample point cannot be less than the sum of the distances from this point to the nearest and the second nearest among the neighbouring sample points. According to (3.4.6) this sum is $0.5 + 0.75 = 2 \cdot 0.625$.

The upper bound corresponds to a serpentine road of the same kind as the one dealt with in the earlier cases of random sampling. It is found in the following way. The plane is divided into strips that have a width of w and are parallel to the x-axis. Let

$$\ldots (x_{-1}, y_{-1}), \; (x_0, y_0), \; (x_1, y_1), \ldots$$

be the sample points in one such strip ordered such that

$$\ldots \leqslant x_{-1} \leqslant x_0 \leqslant x_1 \leqslant \ldots$$

The average distance between two successive points in this sequence is seen to be

$$2 \int_0^\infty \exp(-xw)\, dx \int_0^w (1 - y/w) \sqrt{x^2 + y^2}\, dy \tag{5.7.5}$$

By numerical integration it can be shown that the minimum of (5) is 0.9212. It is attained for w close to $\sqrt{3}$. In this way the right hand side of (4) is obtained, as well as a rule for a path through the random points.

Judging from sampling experiments, carried out by the author, it seems probable that $E(L)$ is close to 0.70. Therefore, it should usually be easy by occasional deviations from the rule to get an average L not insignificantly lower than 0.9212. The value 0.9212 has been entered in table 13, however.

Values for other sampling intensities than 1 can easily be obtained from table 13. If the ratio is changed to 1 point per area A, the averages of the table should be multiplied by $\sqrt{A}$.

5.8. Comparisons between some cases of stratified and systematic sampling

The sections 5.1—5.6 give some information about the ranking of different point sampling designs. We shall now give some indications about the quantitative measurement of the merits of the schemes. The discussion will be confined to the systematic and stratified designs and to the case of an isotropic exponential correlation. It will also be supposed that the region surveyed extends to infinity in all directions.

We shall express the comparative merits of two sampling schemes by aid of the number of sample points necessary in each scheme to obtain a prescribed precision. The ratio between these two numbers is a *relative efficiency* (cf. Kendall & Buckland 1957). As a complement, the corresponding travel distances will also be evaluated.

Let us first consider the systematic design. With the cov.f. $\exp(-hv)$ and the 1×1 square lattice of sample points, the variance per point is found from (5.4.3) or (5.4.4). Write $V_1(h)$ for this variance. Some values of V_1 can be obtained from table 9: $V_1(0.1) = 0.0229$, $V_1(0.5) = 0.114$, etc.

In the general case of an $a \times a$ square grid the *variance per unit area* is

$$a^2 V_1(ah)$$

The meaning of the term variance per unit area should be clear from the following relation: If σ^2 is the variance per unit area with a certain sampling plan, then the estimate of the mean of a region of area A has the variance σ^2/A, if border effects still are negligible. We now introduce the function

$$\psi_1(x) = x^2 V_1(x) \tag{5.8.1}$$

Thus, when the cov.f. is $\exp(-hv)$ and an $a \times a$ square grid is used the variance per unit area is

$$\psi_1(ah)/h^2$$

Further let $V_2(h)$ be the variance per sample point in a stratified random sampling with the stratum equal to the 1×1 square. In the general case of k sample points per stratum of size $a \times a$ the variance per unit area is (cf. 5.2.3)

$$a^2 V_2(ah)/k$$

It is convenient to use a function ψ_2, corresponding to (1) and defined by the equation

$$\psi_2(x) = x^2 V_2(x) \tag{5.8.2}$$

The variance per unit area in the preceding case can now be written

$$\psi_2(ah)/h^2 k$$

The scanty information about the functions (1) and (2) which can be found from table 9 has been supplemented by the computation of some additional values. Thus the figures of table 14 have been obtained. For the numerical methods used, see § 5.2 and § 5.4.

Table 14 can now be used as shown in the following example. Suppose that the cov.f. is $\exp(-hv)$ and that a variance of 1 per unit area is prescribed. This precision can be obtained by an unrestricted random sample with intensity 1.

Table 14. Values of $\psi_1(h)$ and $\psi_2(h)$.

h	$\psi_1(h)$	$\psi_2(h)$
0.1	0.0002288	0.0005051
0.2	0.001829	0.003917
0.3	0.006166	0.01282
0.4	0.01459	0.02948
0.5	0.02849	0.05590
0.6	0.04904	0.09381
0.7	0.07767	0.1447
0.8	0.1156	0.2100
0.9	0.1640	0.2908
1.0	0.2240	0.3881
1.25	0.4325	0.7093
1.5	0.7372	1.149
1.75	1.152	1.716
2	1.689	2.414
2.5	3.167	4.215
3	5.223	6.565
4	11.15	12.91
6	30.10	32.04
8	57.80	59.53

If a systematic sample with an $a \times a$ square grid is used, a must be determined from the equation

$$\psi_1 (ah)/h^2 = 1$$

If e.g. $h = 0.25$, graphical interpolation gives the solution $a = 2.61$; i.e. one sample point per 6.8 units of area, or 0.15 points per unit area. In a stratified sample with k points in each stratum of size $A \times A$, A shall be chosen so that

$$\psi_2 (Ah)/h^2k = 1$$

With $h = 0.25$, $k = 1$, we find graphically the approximate solution $A = 2.08$, or 1 point per 4.3 units of area; 0.23 points per unit area. Thus the efficiency of the stratified sampling is 0.8 ($= 0.15/0.23$) when compared with the systematic design. It should be noted in this case that the relative efficiency depends not only on the structure of the stochastic process but also on the precision required.

Computations of the required number of sample points have been carried out for four values of h: 1/16, 1/4, 1, and 4. Some other sampling schemes have been included in addition to those mentioned. However, in these cases, very few values of the functions corresponding to ψ_1 and ψ_2 have been available. Approximate additional values have been obtained by graphical interpolation where the curves of ψ_1 and ψ_2 have been useful in governing the course of the remaining graphs.

The results are shown in table 15. For each value of h and each sample plan, the table gives the number of points (n) per unit area, which is needed to

Table 15. Number of points (n) and travel distance (t) per unit area in different sample schemes. Covariance function $\exp(-hv)$.

All schemes in the same column give estimates with equal precision. The table refers to the limiting case when the region surveyed extends to infinity in all directions. For a rectangular grid and a rectangular stratum the ratio between the lengths of the sides of the rectangle is denoted α.

Sampling scheme		$h \to 0$	$h = 1/16$	$h = 1/4$	$h = 1$	$h = 4$	$h \to \infty$
Systematic							
Square grid, $\alpha = 1$.	n	0.58	0.06	0.15	0.35	0.75	1.00
	t	0.76	0.24	0.38	0.60	0.86	1.00
Rectangular grid							
$\alpha = 4$.............	n	1.06	0.11	0.27	0.60	0.90	1.00
	t	0.52	0.16	0.26	0.39	0.48	0.50
$\alpha = 16$............	n	3.85	0.39	1.00	2.10	2.30	1.00
	t	0.49	0.16	0.25	0.36	0.38	0.25
$\alpha = 64$............	n	15.2	1.6	3.8	7.7	8.3	1.00
	t	0.49	0.16	0.24	0.35	0.36	0.125
Stratified random, 1 point per stratum							
Square stratum . . .	n	1.00	0.10	0.23	0.49	0.83	1.00
	t	1.09	0.34	0.52	0.76	0.99	1.09
Rectangular stratum							
$\alpha = 2$.............	n	1.06	0.10	0.24	0.51	0.83	1.00
	t	0.93	0.29	0.45	0.65	0.83	0.91
$\alpha = 4$.............	n	1.23	0.12	0.27	0.55	0.85	1.00
	t	0.99	0.31	0.47	0.67	0.83	0.90
Stratified random, 2 points per stratum							
Square stratum . . .	n	1.26	0.12	0.28	0.58	0.90	1.00
	t	1.04	0.32	0.49	0.70	0.88	0.93
Unrestricted random	n	∞	1.00	1.00	1.00	1.00	1.00
	t	∞	0.92	0.92	0.92	0.92	0.92

obtain the same precision as that in unrestricted random sampling with one point per unit area. Further, the table gives the length (t) per unit area of the corresponding path through the sample points. Owing to the crude method of computation, the values of table 15 must be regarded as approximate.

In addition to the four h-values mentioned, calculations have also been carried out for the two limiting cases: $h \to 0$ and $h \to \infty$. When h tends to infinity, the correlation tends to be inappreciable for any finite distance and all sample schemes become equivalent with respect to the precision. When h approaches 0, all plans involving stratification and all systematic schemes tend to become infinitely efficient in relation to the simple random sampling. In this case the values of n have been calculated from series developments in powers of h of the corresponding variances. Taking the two previously treated cases as examples, we find from (5.4.4) and (5.4.5)

$$\psi_1(x) = x^3 \pi^{-2} \sum_{\mu=1}^{\infty} \sum_{\nu=0}^{\infty} (\mu^2 + \nu^2)^{-3/2} + \ldots = 0.2287\, x^3 + \ldots \tag{5.8.3}$$

and from (5.2.6)

$$\psi_2(x) = 0.5214\, x^3 + \ldots \tag{5.8.4}$$

The constants in (3) and the corresponding expressions for other systematic schemes have been obtained by extrapolating the computed values of $\sigma_p{}^2/h$ to $h = 0$, cf. text after (5.4.6). The constants in (4) and analogous expressions in other cases are found from table 5 of § 5.2. In the present case ($h \to 0$) the values of n have been standardized by taking $n = 1$ for the stratified random sample with one point per each 1×1 square stratum. The corresponding systematic sample with an $a \times a$ grid is therefore determined by aid of (3) and (4) as follows

$$\frac{h^3\, 0.5214}{h^2} = \frac{a^3 h^3\, 0.2287}{h^2}$$

or

$$a = \sqrt[3]{0.5214/0.2287} = 1.31. \quad \text{Hence}$$

$$n = a^{-2} = 0.58$$

points are required in the systematic sampling. Analogous computations have given the remaining values of n.

It may be noticed that the value 0.58 just mentioned represents the lowest ratio between the efficiencies of the two sampling designs in question (i.e. stratified sampling with one point per square stratum and systematic sampling of points in a square grid) that can be obtained with a cor.f. of the form $\exp(-hv)$. Similarly, the lower bound of the ratio of the efficiencies of the stratified samples with two points per square and one point per square respectively is $1/1.23 = 2^{-1/3} = 0.79$.

The very few examples of table 15 cannot give any definite guidance as to which scheme gives the required precision with a minimum of travelling. It may be surmised, however, that some rectangular lattice should be near the optimum. If the correlation is appreciable only at very short distances (short in relation to the distance between sample points), then a rectangle with high value of α should be chosen, as illustrated by the columns pertaining to large values of h. The conclusion is the same if the error of measurement or a similar chaotic component is responsible for a considerable part of the variation. In the opposite case (low values of h) it seems not to matter very much which value of α is chosen as long as α is not too near 1. According to the indications obtained by some rough calculations it may be prescribed that α should never be smaller than 3. The problem will be discussed from a slightly different point of view in § 6.11.

As to the random designs, it is seen that the scheme with square stratum and one sample point per stratum always gives longer travel distances than the other three stratified designs. Methods of multi-stage sampling, cluster sampling, etc. may provide considerable reduction of the travel distance in the cases represented by high h-values. It seems hardly possible, however, to construct a random design which would be superior to the best systematic scheme of the rectangular type under the present assumption of exponential correlation.

Chapter 6. Various problems in sample surveys

6.1. Introduction

The sampling schemes which can be considered when information about plane or spatial processes is sought are not confined to the point sampling methods dealt with in the preceding chapter. In fact, a very large variety of plans involving sampling units of 0—3 dimensions can be considered, and examples of their application can easily be found.

It will not be attempted to examine these other schemes in the detailed manner of Ch. 5. Instead, a number of miscellaneous questions that are associated with different sampling methods will be discussed; each topic will therefore get a very brief treatment. Most of the questions emanate from problems encountered in forest surveys. It seems appropriate to review in this introductory section the background and interrelationship of the questions.

The sampling of points in one dimension has been studied by several authors, see references in § 5.1. The so-called line-surveys in two dimensions can also be treated as one-dimensional samples of points. This is achieved by considering the projection of the plane on a line perpendicular to the direction of the survey lines. The sampling of a three-dimensional manifold (e.g. microscopic studies of tissues) by means of parallel plane sections can also be reduced to a one-dimensional problem in an analogous way. References on the line-survey are found in Matérn (1947). An example of the three-dimensional problem is found in Block (1948).

A brief study of the point sampling in R_1 (§ 6.2) shows that the relative merit of the systematic scheme is somewhat less conspicuous than in R_2.

If systematic sampling is applied to a time-series, the possibility of periodic variation creates special problems. Such problems must be considered for example if the random selection of time points in the "ratio delay" method of time studies is to be replaced by a systematic selection. (Cf. Kilander 1957, p. 22. The "ratio delay" or "snap reading" method consists in recording at random instants the state of an activity, thus providing estimates of the average percentage of time the activity is in the different states, see Tippet 1934 and Barnes 1957.) Some illustrations of this question are also given in § 6.2.

It is usually possible to estimate the sampling error in random sampling from the sampling results themselves. However, there are cases which present difficulties. The stratified sampling with selection of only one unit per stratum is one such case. If this scheme is used when samples are taken from a stationary time series or a stationary spatial series, it will be possible, however, to find methods of estimating the sampling error that impose no seriously limitating assumptions on the phenomenon under survey. This is discussed in the one-dimensional case (§ 6.3). The methods can immediately be extended to higher dimensions.

The sampling scheme of § 6.3 is used in § 6.4 to illustrate a situation where it is important to have accurate estimators of the sampling error. This occurs in multi-phase sampling when regression methods are used to estimate the population means. To take an example from forestry, we may want to combine a large-scale sample of eye-estimates or estimates from aerial maps with a small sub-sample of measurements in the field (see Matérn 1947, Ch. VII). In this case the method of estimating the error affects the primary estimates of the population means directly.

It is particularly in conjunction with systematic sampling that the estimation of the sampling error presents difficulties. Valid estimates of the sampling error can be obtained from the sample itself if some special a priori assumptions on the structure of the population are made. In other cases recourse can be had to supplementary data for the specific purpose of obtaining information about the sampling errors (cf. Yates 1948). However, a large number of methods of estimating the sampling error from the data of a systematic survey have been suggested. They are usually based on quadratic forms in the observations. We shall compare different such forms using the same approach as in Ch. 5. Thus we shall study the average performance of the formulas in an ensemble of realizations of a stationary process with decreasing correlation function.

Section 6.5 treats the systematic sample in R_1 with the before-mentioned restriction to the case of decreasing correlation. Nevertheless, it seems impossible to find any general method of obtaining unbiased or even nearly unbiased estimates of the variance of the sampling error. The methods seem to have a general tendency to overestimate the variance.

However, very accurate a posteriori estimates of the sampling error are not always required. The need for information about the precision is often more urgent in the planning stage. We may therefore want to utilize the data from a sample to obtain information about the precision that may be expected in future surveys of similar populations. Data from a systematic sample can usually furnish estimates of the precision of systematic samples of lower intensity than that of the original survey. This is illustrated in the papers by Langsæter (1932) and Östlind (1932) referred to in § 5.1. However, the determination of the "degrees of freedom" among the sub-sample means presents some difficulties which will also be discussed in § 6.5.

The following section (6.6) deals with the case of a systematic sample in R_2. The section gives some examples supplementary to those found in Matérn (1947). If the correlation is isotropic and decreasing with distance, there are (except in the case of some special designs) methods of estimating the precision which do not give any serious bias. Some of the methods are more "robust" than others in the sense that no serious underestimation of the sampling error occurs when the methods are applied to cases that show some deviations from isotropy.

Like the discussion in the author's earlier paper on the problem of estimating the error, the treatment so far (§§ 6.3—6.6) is concerned with the limiting case when the border effects are negligible. Methods to handle the border effects are described in § 6.7.

The following three sections deal with problems associated with the sampling schemes used in the third national forest survey of Sweden. Some features of this design will now be reviewed. For details the reader is referred to Hagberg (1957).

The primary sampling units, so-called *tracts*, consist of squares with a side-length of 1—2 km. A systematic sample of tracts is taken annually. Observations on land-use classes, forest site-classes, etc. are made along the periphery of the tract. Circular plots of radius 6.64 meters constitute secondary sampling units. There are usually four *sample plots* of this kind on every side of the square. All trees on the sample plots are calipered. A certain number of *sample trees* are measured with respect to various features. Additional sample plots, so-called *stump plots*, are located 100 meters apart on the periphery of the tract. In these plots stumps after recent fellings are recorded. The scheme is of a rotational type, and provides for a re-survey of the same tracts at ten-year intervals. Thus the tracts surveyed in ten years constitute ten different samples. Jointly they form a systematic sample with a sampling intensity ten times higher than that of one year's tracts.

The size of a survey tract is different for various regions of the country. It is chosen so as to enable the completion of one tract per day. The reason for making the tract a closed curve is that the crew then will return to its starting point, usually an intersection with the road system. A problem which arises in this context is the shape that a closed figure should have to give efficient estimates. Some comments on this question are given in § 6.8.

Two different schemes for locating the sample plots along the periphery of a square are compared in § 6.9. The case of sample plots along the sides of an octagonal tract is included in the comparisons.

As already mentioned, the size of a tract in the Swedish survey should correspond to one day's work. Yet, it is of interest to study the influence of a slight change of tract size on the precision. This problem is treated in § 6.10.

In 6.10, as in most of the other sections, the stationary process with exponential correlation is chosen for numerical illustrations. The computation technique used in 6.10 can also be applied to illustrate the difference in efficiency between sample plot surveys and strip surveys. Other illustrations of this problem are afforded by the calculations of § 5.4.

If the survey crew proceeds along straight lines between the sample plots, little additional time is required to record land use classes, site classes, etc. along these lines. However, some additional costs are incurred in the field work and in the processing of the data. The extra observations are warranted if corresponding estimates are distinctly better than those which can be based solely on the plots. A comparison between plots and lines is found in § 6.11.

The important question of choosing the size and shape of the sample plot in a forest survey will not be discussed, since investigations of this question are initiated at the Forest Research Institute. Some comments, however, shall be made in § 6.12 on the empirically deduced "Fairfield Smith's law" about the dependence of the variance on the size of a plot. It may be remarked in this context that the ranking of different geometric figures according to their performance as sample plots must be exactly opposite to the ranking according to their properties as strata in sampling of points (§ 5.2).

6.2. Point sampling in R_1

The problem discussed in the present section concerns the estimation of a mean value

$$z = (1/n) \int_0^n z(x)\, dx \qquad (6.2.1)$$

by means of the arithmetic mean $\bar{z}$ of a number of observations made in sample points $(x_1, x_2, \ldots)$ chosen in the interval.

The case when $z(x)$ is a sample function of a real stationary process with exponential correlation function has been treated by several authors, cf. Cochran (1946) and Yates (1948). For convenience the relevant expressions for the variance will be given here. If n equidistant sample points are selected (x_1 at random between 0 and 1, $x_i = x_1 + i - 1$), the variance in the general stationary case is found from

$$nE\{[\bar{z} - z]^2\} = \sum_{-n}^{n} [1 - |j|/n]\, c(j) - \int_{-n}^{n} [1 - |x|/n]\, c(x)\, dx \qquad (6.2.2)$$

In the limit when $n \to \infty$, (2) can be replaced by

$$\sum_{-\infty}^{\infty} c(j) - \int_{-\infty}^{\infty} c(x)\,dx \tag{6.2.3}$$

When $c(x)$ equals $\exp(-|hx|)$, we obtain the special case

$$\coth(h/2) - 2/h \tag{6.2.4}$$

In a stratified sampling where each stratum is an interval of length a, the variance per sample point is

$$c(0) - (1/a) \int_{-a}^{a} c(x)\,[1 - |x|/a]\,dx \tag{6.2.5}$$

in the general case, and

$$1 - 2/ah + 2(1 - e^{-ah})/a^2h^2 \tag{6.2.6}$$

in the exponential case. It is here supposed that n is a multiple of a.

We can now make a comparison between different sample schemes in the manner of § 5.8. Table 16 shows the number of sample points per unit length needed under different conditions to obtain a required precision. Besides the unrestricted random sampling the sampling plans included are the systematic scheme, and the stratified plans with one and two sample points per stratum, respectively. With respect to the systematic scheme, the table refers to the limiting case when the variance is given by (4). The table shows marked differences in efficiency between the three sample methods for small values of h. In these cases it is further seen that the difference in efficiency between the plan with 2 points per stratum and the one with 1 point per stratum approximately equals the difference between the latter scheme and the systematic design. This agrees with empirical evidence (see Yates 1948, p. 372). When the corresponding comparisons were made in R_2 (table 15), it was found that the

Table 16. Number of points per unit length in three different sampling schemes. Covariance function $\exp(-hv)$.

The schemes in the same column give estimates with equal precision. The table refers to the limiting case where the number of sample points tends to infinity.

Sampling scheme	$h \to 0$	$h = 1/16$	$h = 1/4$	$h = 1$	$h = 4$	$h \to \infty$
Systematic.........	0.707	0.102	0.202	0.388	0.670	1.000
Stratified random						
1 point per stratum	1.000	0.137	0.258	0.457	0.708	1.000
2 points per stratum	1.414	0.189	0.347	0.582	0.817	1.000
Unrestricted random	∞	1.000	1.000	1.000	1.000	1.000

gain with the square grid type of systematic sampling over stratified random sampling with one point per stratum is greater than the gain of the latter plan over the stratified sampling with two points per stratum.

The exponential correlation is a special case of formula (3.5.18) which belongs to a model of the development of a system alternating between two states. The model also contains cases in which the cov.f. is non-monotonic. In degenerate cases it may even be strictly periodic. The general expression is a sum of damped sine waves, the damping function of which is exponential. Thus, the terms are of the type

$$\exp(-\alpha v) \cos(\beta v)$$

The corresponding component in the variance per point in systematic sampling with sampling interval a is seen to be $f_1(a\alpha, a\beta)$, where

$$f_1(\alpha, \beta) = \frac{\sinh \alpha}{\cosh \alpha - \cos \beta} - \frac{2\alpha}{\alpha^2 + \beta^2}$$

For $\alpha = h$, $\beta = 0$, we obtain (4) as a special case. The variance within a stratum of length a is $f_2(a\alpha, a\beta)$, where

$$f_2(\alpha, \beta) = 1 - \frac{\cos \theta}{\beta} - \frac{2}{\alpha^2 + \beta^2} [\sin \theta - \exp(-\alpha) \sin(\theta + \beta)]$$

wherein

$$\sin \theta = (\beta^2 - \alpha^2) / (\beta^2 + \alpha^2)$$

Putting $\alpha = h$, $\beta = 0$, we get the special case (6). By these formulas it is possible to compute the precision in sampling a realization of any process with cov.f. of the type (3.5.18).

Variances have been computed for the systematic sampling and the stratified random sampling with one sample point per interval in two special cases, namely $m = n = 2$ and $m = n = 3$ in (3.5.18). With $m = n = 2$ the variance in systematic sampling was found to be between 50.0 and 100.7 per cent of the variance in stratified sampling with the same sampling intensity. The cov.f. is in this case

$$\exp(-\alpha v) \cos(\alpha v)$$

With $m = n = 3$, the cov.f. is

$$[\exp(-2\alpha v) + 8 \exp(-\alpha v/2) \cos(\alpha v \sqrt{3}/2)] / 36$$

In this case the variance of the systematic sample ranges from 50.0 to 109.1 per cent of the variance of the stratified sampling of the same sampling intensity.

In these two examples, the systematic sample is less efficient than the stratified sample with one point per stratum for some values of the sampling ratio.

The efficiency of the systematic sampling attains its lowest value when the sampling interval is approximately equal to the average duration of a whole cycle in the process (including one interval in each state), as is also intuitively clear. The highest efficiency is obtained in the limit when the sampling ratio approaches infinity.

The examples seem to indicate that a considerable loss of precision by applying a systematic design appears only when the stochastic process comes rather close to the case of strict periodicity.

6.3. Estimating the sampling error from the data of a stratified sample with one sampling unit per stratum

Let $z(x)$ be a realization of a real stationary process in R_1. As an estimate of the mean value (6.2.1) we use the arithmetic mean of a sample of n values

$$z_1, z_2, \ldots, z_n$$

Here z_i is written for $z(x_i)$, where x_i is supposed to be chosen at random (uniform distribution) in the interval $i-1<x<i$. All the x_i's are assumed to be independent.

The variance per sample point is given by (6.2.5) with $a = 1$. We use here σ_a^2 to denote the variance per point within a stratum of length a.

A common device used to estimate a variance in the case of one sampling unit per stratum would here imply that we form the average of a number of expressions of the type

$$\frac{1}{2}(z_{i+1}-z_i)^2 \tag{6.3.1}$$

as an estimate of σ_1^2. The expectation of (1) is

$$\sigma'^2 = 2\sigma_2^2 - \sigma_1^2 \tag{6.3.2}$$

If the cor.f. is decreasing, σ' is not only greater than σ_1 but also greater than σ_2. In the case $c(v) = \exp(-hv)$ with small h, we have approximately

$$\sigma_a^2 = ah/3 \qquad \sigma'^2 = h \tag{6.3.3}$$

Thus (2) may give a considerable overestimation.

However, as indicated by Yates (1948, p. 376), it is possible to form a consistent estimate of σ_1^2 from the data. Let t_i be defined by

$$t_i = \begin{cases} 0 & x_{i+1}-x_i > 1 \\ \dfrac{2}{x_{i+1}-x_i} - 2 & \text{otherwise} \end{cases}$$

and form the ratio

$$s^2 = \sum_{1}^{n-1} t_i (z_{i+1} - z_i)^2 / \sum_{1}^{n-1} t_i \tag{6.3.4}$$

It is immediately seen that the expectation of the numerator of (4) is σ_1^2 times that of the denominator. However, the corresponding variances are infinite, which indicates that (4) requires a very high value of n to become an effective estimate. It may therefore be advisable to replace t_i by a step function. Thus an approximation such as

$$(5\, s_0^2 + 3\, s_1^2 + s_2^2)/9 \tag{6.3.5}$$

may be suggested. Here s_j^2 denotes the average of expressions (1) for which

$$j/3 < x_{i+1} - x_i < (j+1)/3 \tag{6.3.6}$$

When $c(v) = \exp(-hv)$ with a small h, the expectation of (5) is approximately

$$158\, h/405 = 0.390\, h$$

This refers to the conditional expectation of (5) when differences are found in each one of the three intervals (6).

Similar methods are applicable in spaces of higher dimensions, and in fact whenever use is made of a continuous variable of stratification (on this concept see Dalenius 1957, pp. 159 ff.). In other cases, sharp boundaries may exist between the strata. No information of within stratum variance can then be obtained from observations on the variation between units in different strata.

It should be emphasized that the use of formulas such as (4) or (5) is not confined to the case of strict stationarity. The formulas can be derived under an assumption of "*local stationarity*" which implies that a stationary model is sufficiently accurate for each section of the axis that equals the length of two strata. Similar extensions of the validity of formulas for estimating the sampling error may also be made in the cases dealt with in the following sections of this chapter, cf. Matérn (1947, pp. 56—57, 127).

The approach to the problem of error estimation which is basic to this and other sections of Ch. 6 can be expressed, somewhat vaguely, in the following way. The unknown variance is a certain functional of the cov.f. $c(v)$. The estimate of the variance has an expectation which is also a functional of $c(v)$. In order that the estimate shall be "nearly unbiased" for a wide class of functions $c(v)$, the two functionals must show a close resemblance.

6.4. A digression on two-phase sampling

Let z_1 and z_2 be realizations of two stationarily correlated processes in R_1 (cf. § 2.2). Let the mean values be m_1 and m_2 and the autocovariance functions c_1 and c_2 respectively. Denote further the cross covariance function by c_{12}.

We shall consider the problem of estimating the mean value of $z_2(x)$ in the interval $0 < x < n$ by means of a two-phase sample: a large sample of x-values for which z_1 is known and a small subsample for which also the values of z_2 are determined. To avoid unnecessary complications we shall deal with the limiting case when $z_1(x)$ is known for all values of x in the interval. We are then concerned with the estimation of

$$z_2 = (1/n) \int_0^n z_2(x)\, dx \tag{6.4.1}$$

by an expression of the form

$$\tilde{z}_2 = \bar{z}_2 + b\,[(1/n) \int_0^n z_1(x)\, dx - \bar{z}_1] \tag{6.4.2}$$

where $\bar{z}_1$ and $\bar{z}_2$ are the sample means. The sample is assumed to consist of n points, chosen by the stratified plan discussed in the preceding section.

The precision of (2) as an estimate of (1) depends on the procedure used for the computation of b from the sample. We assume that, with this procedure, b has the expectation β. As in earlier sections, the expectation is here taken both over realizations of the process and over different outcomes of the selection of sample points. Supposing further that n is large, we approximate the variance of (2) as σ^2/n, where (see 6.2.5)

$$\sigma^2 = c_2(0) - 2\beta c_{12}(0) + \beta^2 c_1(0) - \\ - \int_0^1 [c_2(v) - 2\beta c_{12}(v) + \beta^2 c_1(v)]\, 2(1 - v)\, dv \tag{6.4.3}$$

Consider then the case when the estimation of the error is based on the variance between sampling units in two adjacent strata. Correspondingly, b is found by minimizing an expression such as

$$\sum_1^n \{z_2(x_{i+1}) - z_2(x_i) - b\,[z_1(x_{i+1}) - z_1(x_i)]\}^2 \tag{6.4.4}$$

(cf. 6.3.1). Using (6.3.2) and (6.2.5) we get approximately

$$\beta = \frac{c_{12}(0) - \int_0^2 c_{12}(v) f(v)\, dv}{c_1(0) - \int_0^2 c_1(v) f(v)\, dv} \tag{6.4.5}$$

with

$$f(v) = \begin{cases} v & 0 < v < 1 \\ 2 - v & 1 < v < 2 \end{cases}$$

The corresponding estimate of the variance of (2) is

$$T/2n \tag{6.4.6}$$

where T is the minimum of (4). (The divisor may be replaced by $2n-2$.)

A numerical example shows that this may lead to an estimate $\tilde{z}_2$ that has a lower efficiency than that of the sample mean $\bar{z}_2$.

Assume that for all $v > 0$

$$c_1(v) = c_2(v) = c_{12}(v) = \exp(-0.2v)$$

and that

$$c_2(0) = c_{12}(0) = 1 \qquad c_1(0) = 1.6$$

This corresponds to a situation where z_1 equals z_2 with a superimposed purely random error of observation and possibly an additive bias, independent of x. Suppose further that z_2 is observed in such a large sample that the formulas in the limiting case are applicable and that n is sufficiently large to admit the use of the "large-sample approximations" mentioned above.

In practical survey, this detailed information about the structure of the two-dimensional process z_1, z_2 is not available. We therefore assume that b is estimated by minimizing (4) and that (6) is used for estimating the error. The large-sample approximations give

$$\beta = 0.2293 \qquad T/2n = 0.1376/n$$

The corresponding estimate of the variance of $\tilde{z}_2$ is approximately $0.1785/n$.

Using unbiased estimates of the variance it is found that $\tilde{z}_2$ and $\bar{z}_2$ have the approximate variances

$$0.0692/n \quad \text{and} \quad 0.0635/n$$

respectively. Thus, instead of the apparent decrease in error variance from $0.1785/n$ to $0.1376/n$, the use of the concomitant variate has resulted in a real loss of information concerning the value of (1).

If instead (3) is minimized, β is approximately 0.0956. The corresponding estimate of (1) has the asymptotic variance

$$0.0574/n$$

Although this numerical example is rather extreme, we may conclude that it is especially important to have "practically unbiased" estimators of the sampling errors in multi-phase sampling. This may call for a very careful examination of the way in which the errors are estimated from the data of the survey, or for the collection of additional data for the purpose of estimating the error.

It might be added that a detailed description of the use of regression estimates in connection with systematic sampling has been given in Matérn (1947, Ch. VII).

6.5. Estimating the sampling error from the data of a systematic sample in R_1

Let

$$z_1, z_2, \ldots, z_n \qquad (6.5.1)$$

be observations attached to the sample points $x_1, x_2, \ldots, x_n$, where x_1 is chosen at random between 0 and 1, and $x_{i+1} = x_i + 1$. The arithmetic mean of the n observations, $\bar{z}$, is used to estimate the corresponding average of all z's in the interval $(0 < x < n)$, z in (6.2.1). In the following it will be assumed that n is so large that the variance per sample point can be computed with sufficient accuracy from (6.2.3).

We shall now consider non-negative quadratic forms in the observations (1) that are intended to estimate the variance per sample point. Each form of this kind can be written as an average of a number of squares of the type

$$(a_1 z_1 + \ldots + a_n z_n)^2 \qquad (6.5.2)$$

In order to get meaningful estimates the following restrictions must be imposed on the coefficients of (2)

$$a_1 + \ldots + a_n = 0 \qquad (6.5.3)$$

$$a_1^2 + \ldots + a_n^2 = 1 \qquad (6.5.4)$$

While (3) is required to make (2) independent of the mean $E[z(x)]$, (4) is needed to give unbiased estimates when all z's are uncorrelated random variables. Thus, if z contains a purely random, "chaotic", part (see 4.3), the corresponding component in the variance of the sampling error will automatically get an unbiased estimate. The introduction of a purely random component into any numerical example used in this and the following section would therefore tend to reduce the relative amount of bias in any formula that satisfies (4).

When (3) and (4) are satisfied, (2) has the expectation

$$c(0) + 2 \sum_{1}^{n-1} c(j) A_j \qquad (6.5.5)$$

with

$$A_j = a_1 a_{1+j} + a_2 a_{2+j} + \ldots + a_{n-j} a_n \qquad (6.5.6)$$

The following two special cases of (2) will be considered

$$D_k = (\Delta^k z_1)^2 \Big/ \binom{2k}{k} \qquad (6.5.7)$$

$$Y_k = [-z_1 + 2z_2 - 2z_3 + 2z_4 - \ldots + (-1)^k (2z_k - z_{k+1})]^2/(4k-2) \qquad (6.5.8)$$

Several formulas based on squared differences such as (7) have been suggested for strip surveys of forests, see references in Langsæter (1932). Yates (1953, p. 231) has proposed the use of (8) with $k = 8$ ($k = 8$ is a "convenient compromise"). He calls the linear expression in z's appearing in (8) a "balanced difference".

It may first be remarked that D_k and Y_k should have about the same expectation for a high value of k, since

$$\lim_{k\to\infty} E(D_k) = \lim_{k\to\infty} E(Y_k) = c(0) + 2\sum_{1}^{\infty}(-1)^j c(j) \tag{6.5.9}$$

It should be noticed that (9) equals

$$2V_2 - V_1 \tag{6.5.10}$$

where V_a designates the variance per sample point in systematic sampling with a sampling interval of length a. This relation can be derived from the general expression corresponding to (6.2.3) but it is also clear from the following simple consideration. Let y_1 and y_2 be the means of the observations with odd and even numbers, respectively, in (1). Then

$$(n/4)(y_1 - y_2)^2 = (n/2)(y_1 - z)^2 + (n/2)(y_2 - z)^2 - n[(y_1 + y_2)/2 - z]^2$$

Taking expectations on both sides and letting n tend to infinity, (10) follows. (Cf. the analogous case 6.3.2.)

When the cov.f. is decreasing, V_a is an increasing function of a. In this case (9) exceeds not only V_1 but also V_2.

Now (5) is a quadratic form in $a_1, \ldots, a_n$, and its minimum under the restriction (4) equals the smallest eigenvalue. In the case of exponential correlation, $c(v) = \exp(-hv)$, it is seen that all eigenvalues exceed

$$\tanh(h/2) \tag{6.5.11}$$

see Grenander & Szegö (1958, p. 69). However, (11) is identical with the limiting value (9) in the present case. Thus, the lowest possible expectation of (2) is obtained from (9) or (10) in the exponential case. Also A_1 in (5) gets its minimum value (-1), which is important in the case of a rapidly falling cor.f.

Table 17 shows the expectation of different forms (7) and (8). In the table the true variance according to (6.2.4) and the corresponding values in stratified sampling (with the same sampling intensity and one and two points per stratum, respectively) are included for comparison.

It seems evident that no form of the type considered here can avoid a considerable overestimation of the sampling error in cases where the correlation decreases slowly. When expressions of the type (7) and (8) are used, it is also seen that little is gained by choosing k higher than 6, say. It may further

Table 17. Expectation of various quadratic forms intended as estimates of the variance of systematic samples in R_1. Variance per point in systematic and stratified sampling. Covariance function: $\exp(-hv)$.

	$h=\varepsilon$	$h=0.1$	$h=0.5$	$h=1$	$h=2$	$h=4$
$E(D_1)=E(Y_1)$	ε	0.0952	0.3935	0.6321	0.8647	0.9817
$E(D_2)=E(Y_2)$	$2\,\varepsilon/3$	0.0665	0.3139	0.5546	0.8257	0.9757
$E(D_3)$	$3\,\varepsilon/5$	0.0599	0.2886	0.5244	0.8077	0.9727
$E(D_4)$	$4\,\varepsilon/7$	0.0571	0.2767	0.5088	0.7976	0.9710
$E(D_6)$	$6\,\varepsilon/11$	0.0545	0.2654	0.4931	0.7865	0.9690
$E(Y_4)$	$4\,\varepsilon/7$	0.0570	0.2725	0.4982	0.7876	0.9690
$E(Y_6)$	$6\,\varepsilon/11$	0.0544	0.2620	0.4848	0.7781	0.9672
$E(Y_8)$	$8\,\varepsilon/15$	0.0532	0.2573	0.4787	0.7737	0.9663
$\lim E(Y_k)$	$\varepsilon/2$	0.0500	0.2445	0.4621	0.7616	0.9640
Variance per point						
Systematic sampling.	$\varepsilon/6$	0.0167	0.0830	0.1640	0.3130	0.5373
Stratified sampling						
1 point per stratum..	$\varepsilon/3$	0.0325	0.1478	0.2642	0.4323	0.6227
2 points per stratum.	$2\,\varepsilon/3$	0.0635	0.2642	0.4323	0.6227	0.7812

be remarked that $E(Y_k) \leqslant E(D_k)$ in all cases included in the table. Since Y_k is easier to compute than D_k, it may be concluded that the expression suggested by Yates is "nearly optimal" among the non-negative quadratic forms.

The crucial point in estimating the error is that the observations cannot give information on the behaviour of the cor.f. $r(v)$ for values of v lower than the length of the sampling interval. It seems that Langsæter (1926) is the first author to have recognized this dilemma.

If $k = n-1$ in (8), we get essentially an expression for the variation between two systematic samples with sampling interval 2. It is clear from (10) that this expression cannot provide an unbiased estimate of V_2, contrary to what might be surmised at first glance.

In the more general case, the original sample (1) is subdivided into p systematic samples, each one with a sampling interval p times the original. To avoid complications we assume that $n = pq$, where p and q are integers. Further q will be supposed so large that the end-effects can be neglected. Introducing the average

$$\bar{z}_i = (1/q) \sum_{j=0}^{q-1} z_{i+pj}$$

we form

$$S_p = \sum_{i=1}^{p} q(\bar{z}_i - \bar{z})^2 \qquad (6.5.12)$$

The expectation is (in generalization of formula 10)

$$E(S_p) = pV_p - V_1 \qquad (6.5.13)$$

If the correlation is decreasing with distance, V_1 is less than V_p. Thus

$$p - 1 < E(S_p)/V_p < p$$

(In the exponential case the ratio lies between $p - 1$ and $p - 1/p$.) The "degrees of freedom" by which we shall divide S_p to get an unbiased estimate of V_p are therefore somewhat indeterminate. Especially for small p, expressions of the form

$$S_p/(p - 1)$$

give distinct overestimates. (This applies i.a. to the first values in table 4 of § 4.6.)

A numerical illustration can be obtained by means of an empiric formula found in Östlind (1932). Östlind gives the following expression for the standard error of the estimated cubic volume per hectare in a strip survey of a forest (width of strips 10 meters)

$$0.085\,(L - 10)^{0.75} A^{-0.5} m^{0.6} \tag{6.5.14}$$

Here L is the distance between survey lines in meters; A is the area of the forest in hectares; m is the volume in cubic meters per hectare. This formula has been basic in planning strip-surveys in Swedish forestry.

If $L/10$ is large, the theory of point sampling may be applied. In cases where (14) is valid

$$E(S_p/V_p) = p - p(L - 10)^{1.5}(Lp - 10)^{-1.5} \tag{6.5.15}$$

(L is the distance between lines in the original survey and Lp the distance between lines in the subsamples.) Approximately

$$E(S_p/V_p) = p - p^{-\frac{1}{2}} \tag{6.5.16}$$

It is of a certain interest to apply (15) to the data presented by Östlind, since (14) was derived by somewhat different methods.

Östlind's data are observations from strip surveys of 173 forests in central and southern Sweden. For the volume estimates of the surveys standard errors were computed by a formula of type (7) with $k = 2$. These standard errors were then graduated by means of an expression of type (14). An overall correction factor was applied to the function obtained in this way. For this purpose, the surveys were subdivided into systematic subsamples of lower intensity. The standard error corresponding to a survey of 100 hectares of forest land was estimated from a set of subsamples by the expression

$$(A/100)\,[S_p/(p - 1)]^{\frac{1}{2}} \tag{6.5.17}$$

The estimates (17) were added in a number of classes with varying values of p and L. Estimates from the graduating formula were added in the same way. The ratio between the sums of standard errors computed by the two different methods was used as correction factor.

Some classes were excluded, however, before the final summation took place. It was found that (17) gave rather low values for $p = 2$ and $p = 3$. The corresponding classes were therefore discarded.

It is of interest in this context to know the amount of bias in (17). To obtain a very rough approximation, we assume that the ratio

$$E(\sqrt{S_p})/\sqrt{E(S_p)}$$

is the same as the corresponding ratio for a χ^2 with $p - 1$ degrees of freedom The moment formulas for χ^2 and (16) give

p	2	3	4	5	6	7	8
$E\left[\sqrt{\dfrac{S_p}{(p-1)V_p}}\right]$	0.907	0.975	0.995	1.003	1.006	1.008	1.009

The list covers the values of p used in Östlind's paper. It is seen that the two cases where the bias is appreciable (p = 2 and $p = 3$) correspond to the exclusions made by Östlind.

To avoid some of the complications encountered by Östlind, we may compute the estimated variance per 100 hectares of forest land, with degrees of freedom taken from (15), for every set of subsamples. This can be compared with the variance obtained by squaring (14) with $A = 100$ and m equal to the actual volume per hectare in each particular case.

Variances estimated as indicated above were added for each combination of L and p considered by Östlind (1932, table 3). The sum was then expressed in per cent of the corresponding sum of variances obtained according to (14).

Table 18. Variances estimated as $V_pS_p/E(S_p)$ with $E(S_p)$ from (6.5.15) in per cent of variances according to (6.5.14). Data from Östlind (1932).

Distance between survey lines, meters (Lp)	Sum of degrees of freedom according to (6.5.15)	$V_pS_p/E(S_p)$ in per cent of V_p according to (6.5.14)
200	99.8	59
300	138.0	65
400	86.3	78
450—500	71.0	81
600—750	98.3	63
800—1,200	116.7	100
1,400—4,000	62.6	74
200—4,000	672.7	74

A summary of the results is found in table 18. The percentages have been pooled by using the "degrees of freedom" (from formula 15) as weights.

It is seen that the estimated variances are on the average smaller than the graduated values. This is partly due to a further correction $(1.02)^2$ introduced by Östlind for some technical reasons. Even if this is considered, (14) must be said to include a safety margin, especially for short distances. It seems reasonable that Östlind should get a tendency to give higher overestimation with short sampling intervals than that obtained with long intervals (cf. table 17). Some evidence in the same direction is reported by Hagberg (1958).

The values for large Lp in table 18 are in fact based on rather scanty data. For distances $> 1\,200$ meters, all computations refer to data from only five surveys, which have been subdivided into subsamples in different ways. Therefore, no far-reaching conclusions can be drawn. A priori, it seems not unreasonable, however, that Östlind's formula should overestimate the error for a large distance between lines. It may be argued that the decrease in the cor.f. should be very slow for large distances. This ought to result in a tendency for the sampling error of the survey to increase with L only slightly faster than $L^{0.5}$ and not as fast as the factor $L^{0.75}$ in (14). In conformity with the remarks made earlier, it is reasonable to expect that the sampling error increases more rapidly than $L^{0.75}$ for low values of L.

6.6. Estimating the sampling error from the data of a systematic sample of points in R_2.

The purpose of this section is to discuss error estimation for the sampling schemes presented in § 5.4. (Sections 5.1 and 5.4 contain the basic assumptions and the notation.)

Consider first the case $a = b = 1$, $\varphi = \pi/2$ in (5.4.1), that is the square network of integer lattice points. Assume that the region surveyed is large, so that border effects can be neglected. As in 6.5 only positive semi-definite quadratic forms in the observations will be treated. Such a form can be written as the average of a number of expressions of the type

$$T = \left[\sum_{i=1}^{n} \sum_{j=1}^{n} (-1)^{i+j} a_{ij} z(i,j)\right]^2 \qquad (6.6.1)$$

with (cf. 6.5.3—4)

$$\Sigma\Sigma(-1)^{i+j} a_{ij} = 0 \qquad \Sigma\Sigma a_{ij}^2 = 1$$

The following three cases will be considered.

T_1, obtained from (1) by choosing $n = 2$, $a_{ij} = 1/2$;

T_2, $n = 4$, $a_{11} = a_{14} = a_{41} = a_{44} = 0.05$, $a_{22} = a_{23} = a_{32} = a_{33} = 0.45$, the remaining a's equal 0.15;

T_3, $n = 4$, $a_{11} = a_{14} = a_{41} = a_{44} = 0.1$, $a_{22} = a_{23} = a_{32} = a_{33} = 0.4$, the remaining a's equal 0.2.

T_1 can also be written

$$(1/4)\,[\Delta_i \Delta_j z(1, 1)]^2$$

See further Matérn (1947, p. 90) and Seth (1955, p. 47). Similarly

$$T_2 = (1/400)\,[\Delta_i^3 \Delta_j^3 z(1, 1)]^2$$

T_3, which is found in Yates (1953, p. 231) is a particular case of a general formula corresponding to (6.5.8). If the adjustments of the bordering elements are neglected, the general formula would be

$$n^{-2}\left[\sum_{i=1}^{n} \sum_{j=1}^{n} (-1)^{i+j} z(i, j)\right]^2$$

The squared expression is the difference between sums that pertain to two sets of sample points. Each one of these sets consists of a square network of points with distance $\sqrt{2}$ between next neighbours. We denote by T_4 the limiting case, $n \to \infty$. Similarly, T_5 will denote the limiting case of (6.5.8), $k \to \infty$. It has been shown in 6.5 that among all forms in observations from *one* line of sample points T_5 has certain optimum properties.

Let $z(x, y)$ be a realization of an *isotropic* process, and denote by V_a the variance per sample point in a square system of points with distance a between neighbours.

The previous remark on T_4 means that

$$E(T_4) = 2\,V_{\sqrt{2}} - V_1 \tag{6.6.2}$$

It can be seen that (2) also is the expectation in the limiting case of formulas based on differences in two directions.

Table 19 gives numerical values of the quadratic forms in the case $c(v) = \exp(-hv)$. For comparison, the true variance in systematic sampling and the variances in stratified random sampling (one, respectively two sample points per stratum) are included.

The table shows that all the quadratic forms give a substantial overestimate if h is small. However, a comparison with table 17 indicates that the situation is somewhat more favourable than that in the corresponding one-dimensional case. Table 19 further shows that already the simple form T_2 has smaller bias than T_5, the optimal form among all those which measure the variation in one direction only.

The case when the sample plots are located in the pattern of a "line-plot" survey (short distances between points on the same survey-line, and comparatively long distances between the lines) can be dealt with by the same methods

Table 19. Expectation of various quadratic forms for estimating the variance in systematic sampling in R_2 (sample points in a square network). Corresponding true variance and the variance in stratified sampling with the same sampling intensity. Covariance function: exp(-*hv*).

	$h=\varepsilon$	$h=0.1$	$h=0.5$	$h=1$	$h=2$	$h=4$
$E(T_1)$	0.5858 ε	0.0584	0.2800	0.5074	0.7884	0.9669
$E(T_2)$	0.4663 ε	0.0466	0.2288	0.4343	0.7296	0.9531
$E(T_3)$	0.4540 ε	0.0454	0.2231	0.4250	0.7204	0.9505
$E(T_4)$	0.4183 ε	0.0418	0.2064	0.3968	0.6904	0.9411
$E(T_5)$	0.5000 ε	0.0500	0.2449	0.4621	0.7616	0.9640
Variance per point Systematic sampling with integral lattice points	0.2288 ε	0.0229	0.1138	0.2240	0.4222	0.6970
Stratified sampling 1 point per unit square	0.5214 ε	0.0505	0.2237	0.3881	0.6035	0.8068
2 points per $\sqrt{2}\times\sqrt{2}$ square	0.7374 ε	0.0705	0.2978	0.4922	0.7121	0.8800

as those used for the "line surveys" (the sample consists of parallel equidistant lines).

The error estimation in the case when the sampling units are parallel lines can be treated as a one-dimensional problem, as in the preceding section. Doing so, however, we abstain from using information concerning the short-distance variation, which can be obtained by using short segments of the lines. Formulas based on segments of lines were proposed by Näslund (1939) and several such formulas were examined by Matérn (1947).

A numerical example will now be treated as a supplement to the author's earlier investigation. This example directly pertains to the systematic sampling of points. The following symbols will be used. Points are situated at distance p apart along lines. The distance between neighbouring lines is L. The sample points are of the type (x_i, y_j) with

$$x_i = ip \qquad y_j = jL \tag{6.6.3}$$

A sequence of k sample points on the same line shall be referred to as a *k-section*. The sum of observations from a k-section is written

$$Z(a, b; k) = \sum_{i=0}^{k-1} z(a+ip, b) \tag{6.6.4}$$

Only two types of quadratic forms in Z-values will be treated. The first is an average of squares of the type

$$U = \frac{1}{2nk}\left[\sum_{j=1}^{2n}(-1)^j Z(a+pkj, b; k)\right]^2 \tag{6.6.5}$$

whereas the second expression is based on squares

$$V = \frac{1}{2nk}\left[\sum_{j=1}^{2n}(-1)^j Z(a, b+Lj; k)\right]^2 \qquad (6.6.6)$$

U and V are similar to the squared "balanced differences" proposed by Yates, cf. (6.5.8). While U measures the variation of Z-values along the same line, V is an expression of the variation in a sequence of Z-values belonging to separate lines.

One of the cases dealt with in 5.4, namely $p = 1/4$, $L = 4$, and the exponential cov.f. $\exp(-hv)$, is chosen as a numerical illustration. Confining the study to the limiting case $n \to \infty$, we introduce

$$W_1(k) = \lim_{n\to\infty} E(U) \qquad (6.6.7)$$

$$W_2(k) = \lim_{n\to\infty} E(V) \qquad (6.6.8)$$

The same limiting expectations are obtained if (5) and (6) are replaced by squared differences of order n. In table 20 are found values of W_1 and W_2 for some different k, as well as the variance per point in the sample survey. For the computation of the latter value, see 5.4. In the computation of W_2, the covariance of expressions $Z(a, b; k)$ and $Z(a, b'; k)$ with $b \neq b'$ has been approximated as

$$k^2 \exp(-h|b-b'|)$$

Formulas for the allowable *minimum length of a segment* are given in Matérn (1947). If the segments exceed the minimum length, the corresponding quadratic form cannot have a negative bias in the isotropic case. An application to the present problem gives the following minimum value for k

$$CL/p \qquad (6.6.9)$$

where C depends on the particular quadratic form used. In (5) with $n \to \infty$ we have $C = 2/\pi$ (cf. Matérn 1947, formula 71 b), whereas in (6) the corresponding value is $1/\pi$ for all n (*ibidem*, formula 69 a).

If $L = 4$, $p = 1/4$, the condition for (5) is

$$k \geqslant 10.186 \qquad (6.6.10)$$

For (6)

$$k \geqslant 5.093 \qquad (6.6.11)$$

In agreement with (10) table 20 shows that $W_1(8)$ gives underestimates and that $W_1(12)$ gives overestimates. The bias of $W_1(10)$ is rather small. It is further seen that $W_2(5)$ gives overestimates in all the tabulated cases, while

Table 20. The expectations W_1 and W_2 (6.6.7—8) and the variance per point in the $4 \times 1/4$ systematic sample in R_2. Covariance function: exp(- *hv*).

	$h = 0.1$	$h = 0.5$	$h = 1$	$h = 2$	$h = 4$	$h = 8$
$W_1(6)$	0.158	0.751	1.302	1.719	1.553	1.192
$W_1(8)$	0.274	1.253	1.981	2.195	1.704	1.223
$W_1(10)$	0.422	1.842	2.641	2.537	1.796	1.241
$W_1(12)$	0.603	2.489	3.239	2.784	1.857	1.253
$W_1(16)$	1.058	3.851	4.206	3.104	1.934	1.268
$W_2(3)$	0.527	1.981	2.335	2.052	1.581	1.193
$W_2(4)$	0.668	2.493	2.867	2.387	1.712	1.223
$W_2(5)$	0.793	2.947	3.319	2.641	1.798	1.241
$W_2(6)$	0.903	3.349	3.704	2.838	1.858	1.253
$W_2(8)$	1.079	4.013	4.313	3.116	1.934	1.268
Variance per point in the $4 \times {}^1/_4$ systematic sampling	0.392	1.742	2.578	2.522	1.771	1.215

$W_2(4)$ has a positive bias in some cases and a negative bias in other cases. This is in agreement with (11).

Table 20 will now be used to indicate the result when an error estimation adapted to the isotropic case is used in a case of non-isotropic correlation.

Consider a process $z_1(x, y)$ defined by

$$z_1(x, y) = z(x, \alpha y)$$

where z is isotropic with cov.f. $\exp(-hv)$. The cov.f. of z_1 is

$$\exp\left(-h\sqrt{x^2 + \alpha^2 y^2}\right) \tag{6.6.12}$$

The ratio between the lengths of the axes of the "isocorrelation ellipses" of (12) equals α. Let us further suppose that a realization of z_1 is sampled in the points (x_i, y_j), where

$$x_i = i/4 \qquad y_j = 4j/\alpha \tag{6.6.13}$$

Thus the observations can be written

$$z_1(i/4, 4j/\alpha) = z(i/4, 4j)$$

Hence they correspond to a $4 \times 1/4$ systematic sample of the basic process z. This means that the variance per sample point for $h = 0.1$, 0.5, etc., is the one given in table 20.

However, the observations pertain to z_1 and the sample points (13), not to z. Let us imagine that the sampling is carried out under the false assumption that z_1 is isotropic. Consequently, the error estimation is made by (5) or (6) with the length of the section determined as in the isotropic case. Assuming that (5) is used with a high value of n, k is chosen as the integer closest to

$$32/\pi\alpha = 10.186/\alpha \tag{6.6.14}$$

whereas in (6) the corresponding value is

$$16/\pi\alpha = 5.093/\alpha \tag{6.6.15}$$

It may suffice to consider two numerical examples. Take first the case $\alpha = 5/3$. The correlation is strongest in the direction of the x-axis. From (14) is found $k = 6$, while (15) gives $k = 3$. The expectations of the estimated variances are $W_1(6)$ and $W_2(3)$, respectively. Table 20 shows that $W_1(6)$ always underestimates the variance, $W_2(3)$ gives in some cases an overestimate, in other cases an underestimate. The risk for a substantial underestimation, however, is present only in the case of $W_1(6)$.

A value of $\alpha > 1$ corresponds to the case where the lines of a line-plot survey run parallel to the main topographic direction of the region under survey. However, one usually tries to let the lines cross valleys and ridges perpendicularly. This corresponds to $\alpha < 1$. We consider the example $\alpha = 5/8$. Then (14) and (15) give the values 16 and 8, respectively. Table 20 shows that the corresponding expectations $W_1(16)$ and $W_2(8)$ overestimate the variance. The overestimation is about equal in the two cases.

These examples, and others obtained from table 20, indicate that the formulas based on expressions of type (6) are preferable to those based on (5), since they seem to be less affected by strong deviations from isotropy.

Many other formulas could be considered. We might use k-sections of different lengths in the same formula. Further, expressions of the variation in different directions may be combined, etc. Similarly, for systematic samples in R_3 (for example equidistant plane sections of tissues), a rich variety of expressions for the variation can be obtained from the data.

6.7. Allowance for border effects in estimating the sampling error

The formulas for error estimation given in the preceding sections are derived, like those in Matérn (1947), under the assumption that the border effects are negligible. However, these effects cannot always be neglected. For example, when the total area of a region is estimated by a systematic sample of points, the border effect is the only source of sampling errors, cf. (5.5.5—6) and an example in Strand (1951).

General formulas for the border effect being presented in § 5.5, it may suffice to indicate here by means of an example how the border effects can be considered in the present particular problem.

Take the case of a square grid of sample points in R_2. The sample survey is supposed to pertain to a certain region Q. The following observations are made for each sample point x_i

$$z(x_i) = \text{height above sea-level (e.g.)}$$

$$e(x_i) = \begin{cases} 1 & \text{if } x_i \in Q \\ 0 & \text{otherwise} \end{cases}$$

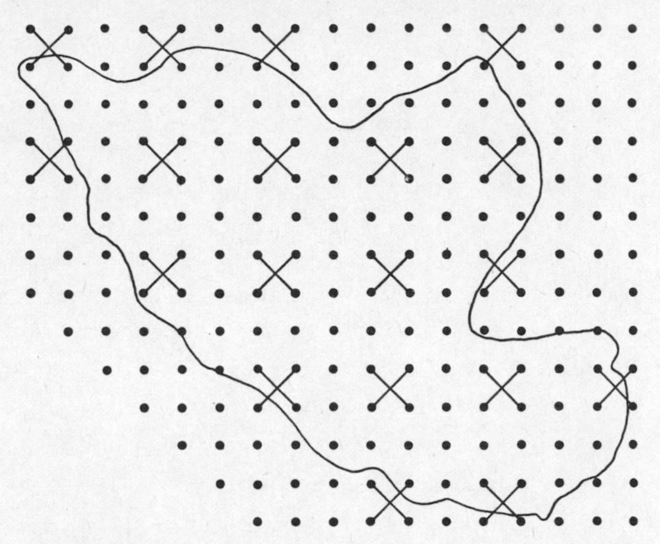

Fig. 13. Sample points used in estimating the error of a systematic survey.

Let the product of z and e be denoted z_1. To estimate the mean $z(Q)$ we use the average

$$\bar{z} = \sum_i z_1(x_i) / \sum_i e(x_i)$$

where the summation can be extended also to the lattice points outside Q.

Assume further for the sake of simplicity that the error-estimation is based on a formula involving groups of only four sample points, viz. the special case T_1 of (6.6.1). The four points of a group form the corners of a square. A certain regular configuration of such quadruplets of sample points is used. Let N be the total number of quadruplets with at least one point inside Q. These quadruplets are indicated by crosses in fig. 13.

Consider the residual

$$z_2(x) = z_1(x) - \bar{z}\,e(x) = e(x)\,[z(x) - \bar{z}]$$

Let x_1, x_2, x_3, and x_4 be the four points in a quadruplet. Put (cf. 6.6.1)

$$T = (1/4)\,[z_2(x_1) - z_2(x_2) - z_2(x_3) + z_2(x_4)]^2$$

Let $T_1, T_2, \ldots, T_N$ be the corresponding expressions for the N quadruplets intersecting Q. Let further q denote the ratio between the number of sample points and the number of quadruplets. This ratio depends on the geometric pattern chosen for the system of quadruplets; in fig. 13, e.g., q is 9. The expression

$$q \sum_{i=1}^{N} T_i \Big/ \Big[\sum_{\nu} e(x_\nu) \Big]^2 \qquad (6.7.1)$$

where the summation in the denominator extends over all sample points with $e \neq 0$, can then be applied as an estimate of $D^2(\bar{z})$.

Since T and z_1 vanish outside Q, both summations in (1) can be regarded as extended over the whole of R_2. In fact, the procedure is essentially a "large sample method" of estimating the error of the ratio $\bar{z}_1/\bar{e}$. The method can readily be adapted also to the more general case of estimating a ratio of the type $Z_3(Q)/Z_4(Q)$, where $Z_3(Q)$ is, for example, the total volume of forest trees in Q and $Z_4(Q)$ is the total forest area in Q. For aspects on the practical computations, see Matérn (1947, pp. 79 ff., 129).

6.8. Linear sampling units ("tracts") in R_2

Let $z(x, y)$ be a realization of an isotropic process in R_2 with cov.f. $c(v)$. Consider also in R_2 a curve C of length P. The average, $z(C)$, is defined as

$$(1/P) \int_C z(x, y)\, ds$$

where ds is the element of arc length measured along C.

The variance is found from

$$D^2[z(C)] = \int_0^\infty c(v) f(v)\, dv \qquad (6.8.1)$$

where $f(v)$ is the frequency function of the distance between two points chosen independently and with uniform distribution over C.

Some special cases should be noted. If C is a circle with radius R

$$f(v) = (1/\pi R)(1 - v^2/4R^2)^{-\frac{1}{2}} \qquad 0 < v < 2R \qquad (6.8.2)$$

If C is a line segment of length P

$$f(v) = 2(P - v)/P^2 \qquad 0 < v < P \qquad (6.8.3)$$

Finally, let C be the contour of a rectangle of size $A \times B$. Then

$$(A + B)^2 f(v) = A\varphi(v/A; B/A) + B\varphi(v/B; A/B) \qquad (6.8.4)$$

with

$$\varphi(x; k) = \begin{cases} 1 - x + \pi x/2 & 0 < x < 1 \\ kx/\sqrt{x^2 - 1} + 2x \arcsin(1/x) - x - \pi x/2 & 1 < x < \sqrt{1 + k^2} \end{cases}$$

Like (2) and (3), $\varphi(x; k)$ shall be understood to vanish outside the intervals for which an expression is given.

When C is the periphery of a closed polygon with all angles obtuse, we find

$$f(v) = \frac{2}{P} + \frac{2v}{P^2} \sum_{\nu=1}^{n} \left(\frac{\pi - A_i}{\sin A_i} - 1 \right) + o(v) \tag{6.8.5}$$

where $A_1, \ldots, A_N$ are the (interior) angles of the polygon.

If C is a closed curve with continuously changing curvature, the frequency function is

$$f(v) = 2/P + (v/2P)^2 \int_0^P [K(s)]^2 ds + o(v^2) \tag{6.8.6}$$

Here $K(s)$ denotes the curvature of C. For a closed curve

$$\int_0^P K(s)\, ds = 2\pi$$

By Schwarz's inequality

$$\int_0^P ds \int_0^P |K(s)|^2 ds \geqslant \left[\int_0^P K(s)\, ds \right]^2 \tag{6.8.7}$$

Hence, whatever the shape of a closed curve of length P, the coefficient of v^2 in (6) cannot fall below π^2/P^3. This minimum is attained when the curvature is constant, i.e. when the curve is a circle.

When the correlation is decreasing rapidly, the course of $f(v)$ in the vicinity of 0 is decisive for the value of (1). It is seen from (5) and (6) that, to a first approximation, the variance is inversely proportional to P, as is also intuitively clear. Comparing closed curves of equal length, we find from (5) and (6) that the curves with continuous curvature give lower variance than the polygons, and that minimum is obtained for the circle. If the restriction to closed contours is removed, it is easily seen that the line segment has the corresponding optimum property in the wider class of continuous curves of given length.

Table 21. Characteristics of eight curves of length 4.

Curve	$\int v f(v) dv$	$\int v^2 f(v) dv$
Line segment	1.333	2.667
Circle, radius $2/\pi$	0.811	0.811
Square 1.0 × 1.0	0.735	0.667
Rectangle 0.8 × 1.2	0.731	0.667
Rectangle 0.6 × 1.4	0.721	0.667
Rectangle 0.4 × 1.6	0.703	0.667
Rectangle 0.2 × 1.8	0.682	0.667
Rectangle 0.0 × 2.0	0.667	0.667

The first two moments can be used to characterize roughly the whole course of $f(v)$. Table 21 shows these characteristics for eight different curves of length 4. The values may give some indication about the performance of the different curves in the case of a slowly decreasing correlation. The curves are shown in fig. 14.

As a further illustration, the variance (1) has been computed for some curves under the assumption that the correlation is exponential. The results are presented in table 22. They are computed by numerical integration.

The cor.f.

$$0.4 \exp(-v) + 0.6 \exp(-12v) \tag{6.8.8}$$

has been used to smooth an empiric correlogram for the distribution of forest land in a Swedish province (Matérn 1947, p. 58). In (8) v is expressed in kilometers. We change the unit to 2 km to obtain variances that refer to figures with a perimeter of 8 km (roughly corresponding to the perimeter of the "tract" used at present in the northern regions of Sweden in the national forest survey). Thus (8) is transformed into

$$0.4 \exp(-2v) + 0.6 \exp(-24v)$$

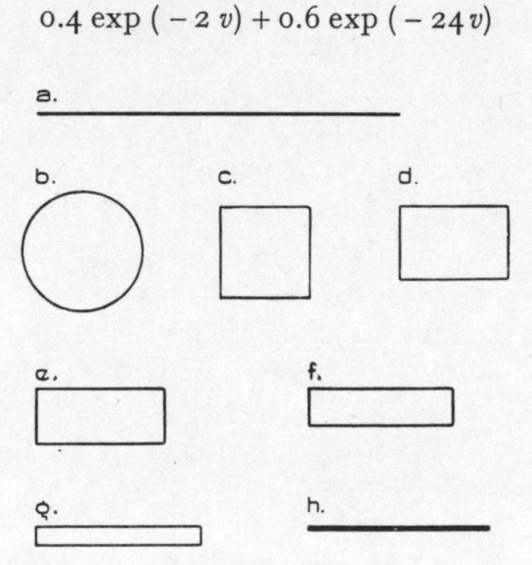

Fig. 14. The 8 curves of table 21.

Table 22. Variance of the mean of observations made along continuous curves of length 4. Covariance function: exp(-*hv*).

Curve	$h = 0.1$	$h = 0.5$	$h = 1$	$h = 2$	$h = 4$
Line segment of length 4	0.879	0.568	0.377	0.219	0.117
Circle, radius $2/\pi$	0.923	0.680	0.482	0.274	0.131
Square 1.0 × 1.0	0.930	0.704	0.512	0.299	0.144
Rectangle 0.5 × 1.5	0.934	0.720	0.536	0.325	0.157
Rectangle 0.0 × 2.0	0.937	0.736	0.568	0.377	0.219

The variances are computed by taking the component corresponding to $\exp(-2v)$ from table 22, while the second component is approximated by means of the asymptotic formulas (5) and (6). Expressing all the variances in per cent of that pertaining to the square, we obtain the following series

Figure	Variance
Line segment (8 km)	75.5
Circle (perimeter 8 km)	92.2
Square (2×2 km)	100.0
Rectangle (1×3 km)	107.7
"Rectangle" (0×4 km)	132.5

It must be noticed that these variances, like those in table 22, are directly applicable only in the case of unrestricted random sampling of a large area. However, the variances can be used to discriminate between different types of figures also in many other designs. Yet, in the case of a systematic selection, the spacing of the sampling units must be such that the covariance between the means of two neighbouring sampling units can be approximated as $c(a)$ where a is the distance between the centers of the two units. Therefore, the variances given here for straight lines are of no use when the sampling units are connected into a continuous chain of segments. Similarly, in stratified sampling, the stratum must be large in comparison with the sampling unit.

As to the inclusion of the circle in the above computations it may be remarked that this figure is hardly a tract feasible in a field survey. However, the data given for the circle may be considered as approximations to those of figures such as the regular hexagon and octagon.

6.9. Locating sample plots on the periphery of a tract

Fig. 15 shows two ways (A, B) of locating 16 equidistant sample points on the periphery of a square tract. When applied in field the two different systems are almost exactly equal as to all items of cost. It might be intuitively felt that system A should give more precise estimates than system B.

In fig. 15 are also shown (C) 16 equidistant sample points on the periphery of a regular octagon of the same perimeter as the square. The reason for considering this case here is that it may be surmised from § 6.8 that the octagon gives more precise estimates than the square when observations are made continuously along the contour. It is therefore of some interest to see if such a conclusion can be drawn also in the case of sample plots.

The observations made on the sample plots can be attached to the centers of the plots. We have then to deal with observations $z(x)$ of the "local integration" type (see 4.4 and 5.1). The observations may also be affected by

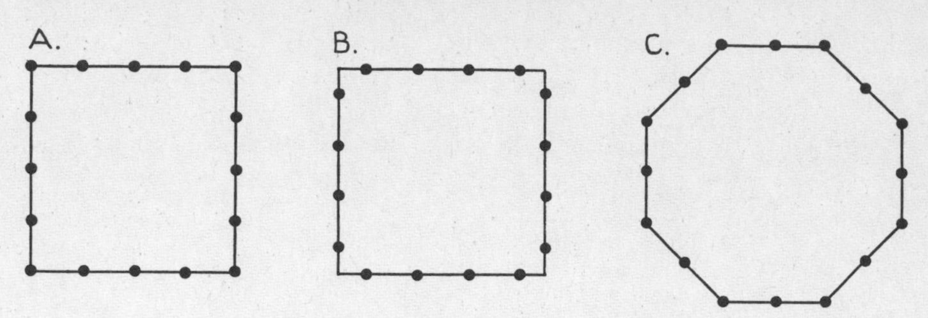

Fig. 15. Locating sample points on the periphery of a square (A, B) and a regular octagon (C).

observational errors. However, from the discussion attached to (5.1.3) it is obviously not necessary to consider the short-distance properties of the variation when our sole object is to *rank* the three methods of locating sample points on a closed contour. Thus, properties connected with the shape and size of the plots may be disregarded. We therefore assume that $z(x)$ is a realization of an isotropic process with smoothly running cov.f., $c(v)$.

The comparisons of the variances per point are based on

$$\frac{1}{16} \sum_i \sum_j c(|x_i - x_j|) \tag{6.9.1}$$

where $|x_i - x_j|$ is the distance between the sample points x_i and x_j, belonging to one or the other of the three systems of fig. 15. It should then be understood that a further component, constant for all the ways of locating the sample plots, should be added. Generally, this component should be decreasing when the size of the sample plot increases.

Values according to (1) have been computed for the three systems in the case $c(v) = \exp(-hv)$. The variances of the systems B and C in per cent of those of A were found to be:

h	0.25	0.5	1	2	4	8
Variance of B	101.5	102.6	103.9	104.3	102.0	100.2
Variance of C	95.8	94.0	94.9	98.3	99.9	100.0

A close comparison between A and B shows that the minimum relative efficiency of B is about 95.7 % in the case of an exponential correlation (the minimum is attained for $h \approx 1.7$) and 91.8 per cent if the correlation is Gaussian, i.e. $c(v) = \exp(-hv^2)$.

Now we also take an example more directly connected with forest survey problems. A correlogram for the volume of stems observed on sample plots was graduated by

$$0.4 \exp(-3v) + 0.6 \exp(-12v) \tag{6.9.2}$$

with the kilometer as unit of length (Matérn 1947, p. 62). If A is once more used as basis, the following variances are obtained when then the plots are located 500 meters apart

A	B	C
100.0	102.6	98.4

Thus, there are small but distinct differences between the three systems. The advantage of the octagonal system, surmised from the investigation in § 6.8, subsists.

6.10. The size of a tract

The precision of a "tract survey" depends i.a. on the size of the tract. Studying this dependence we shall only consider observations from equidistant plots along the sides of a square tract. It is assumed that these plots are located at a spacing of 100 meters, like the stump plots of the current Swedish survey.

For simplicity, assume that the side of the square is always a multiple of the distance between stump plots. Further, suppose that the plots are situated as shown in fig. 15 A. Let n denote the number of plots per side. Thus the total number per tract is $4n$, and the side of the square is $0.1n$ km.

The stump plot observations are now, as in 6.9, represented as values $z(x_1)$, $z(x_2)$, . . . , attached to the centers of the plots. Let us consider the variance *per point*

$$4n D^2 [\Sigma z(x_i)/4n] \tag{6.10.1}$$

This variance has been computed for $n = 12(2)22$, and $c(v) = \exp(-hv)$. The calculations have been made on the electronic computer referred to earlier.

Table 23. Variance per sample point when $4n$ points are located equidistantly on the periphery of a square with the side 0.1 n. Covariance function: $\exp(-hv)$.

	$n=12$	$n=14$	$n=16$	$n=18$	$n=20$	$n=22$	$n=\infty$
$h=1$	21.83	22.76	23.38	23.75	23.96	24.04	20.02
$h=1.5$	15.83	16.01	16.02	15.95	15.83	15.69	13.36
$h=2$	12.02	11.94	11.81	11.66	11.52	11.38	10.03
$h=3$	7.79	7.65	7.53	7.43	7.35	7.29	6.72
$h=4$	5.67	5.57	5.50	5.45	5.41	5.38	5.07
$h=5$	4.45	4.39	4.35	4.32	4.30	4.28	4.08
$h=8$	2.76	2.74	2.73	2.72	2.71	2.70	2.63
$h=10$	2.24	2.23	2.22	2.21	2.21	2.21	2.16
$h=12$	1.91	1.90	1.90	1.89	1.89	1.89	1.86
$h=20$	1.32	1.32	1.32	1.32	1.32	1.32	1.31

The results are presented in table 23. Further, the limiting value when n tends to infinity

$$\coth (h/20) \tag{6.10.2}$$

is shown in the table under the heading $n = \infty$.

Table 23, and some additional calculations, show that the variance per point first is an increasing function of n (for a given h). After passing a maximum, it slowly decreases towards (2).

Consider again the cov.f. (6.8.8) derived from a correlogram for the distribution of forest area. The following variances of the mean of a tract, $\bar{z}$, are obtained on the basis of table 23

n	12	14	16	18	20	22
$D^2(\bar{z})$	139.3	123.9	110.9	100.0	90.7	82.7

The figures are percentages of the variance for $n = 18$. Corresponding calculations for the cov.f. (6.9.2) give

n	12	14	16	18	20	22
$D^2(\bar{z})$	155.5	131.4	113.6	100.0	89.3	80.6

A reduction of the side length by 200 meters gives in both cases an increase in $D^2(\bar{z})$ of 10—20 per cent.

It should be borne in mind (cf. 6.9) that a realistic model should contain expressions for the effect of errors of observation, local integration, etc. The component $\exp(-12\, hv)$ in the covariances (6.8.8) and (6.9.2) may be regarded as a rough description of the influences of these and other factors which are effective over short distances.

The examples suggest that the variance *per plot* is "practically" independent of the size of the tract if the inter-plot distance is fixed. Hence, the variance *per tract* would be inversely proportional to the length of the tract. The same conclusion would be valid also for estimates based on observations made continuously along the contour. However, the conclusion is evidently invalidated if the distance between the plots and the length of the tract side are changed simultaneously.

Finally, it should be noticed that the calculations and the conclusions refer to the case when the tracts are selected by unrestricted random sampling. In many other schemes (cf. the formulas of 5.2—5.4), a term which is approximately independent of the size of the tract should be subtracted. This means that the influence of the size of the tract on the sampling error can be somewhat larger in these cases than that in unrestricted random sampling. However, the formulas will be more complicated and extensive calculations would be needed, for example when the distance between neighbouring tracts in a systematic sample is of the same order as the length of the side of the tract. A similar remark can be made in the case when the tract is of the same order of magnitude as the stratum in stratified sampling. See further the discussion in § 6.8.

6.11. A comparison between strip surveys and plot surveys

One set of observations made in forest surveys consists of records of land-use classes, site-classes, etc. These observations can usually be thought of as referring to the center of the sample plot or to the center of the survey strip. This is applicable also to records of the height above sea-level, the distance to the nearest point on a network of roads or waterways, etc. Treating such observations, we may deal with lines and points instead of strips and plots, respectively.

The distance between lines in a system of equidistant lines is chosen as unit of length. Sample points are located on the lines at a spacing of p, where p is supposed $\leqslant 1$. We assume that the sample points form a rectangular array. The sampling error of the point survey can then be found from computations already utilized in previous sections (5.4 and 6.6). They refer to the limiting case when the size of the region surveyed approaches infinity. The original computations gave the variance per sample point. Now there are $1/p$ sample points per unit length of the lines. Multiplying the previous variances by p, we obtain the variances per unit length. Variances of this kind are presented in table 24 for a series of values of p, and for some cases of the covariances $\exp(-hv)$ and $bv\ K_1(bv)$. The table should be regarded as a byproduct of the earlier investigations. This should explain the gaps in the table and the fact

Table 24. Variance per unit distance in a "line-point" survey. Distance between lines: 1. Distance between points: p.

Covariance function	$p=1$	$p=1/2$	$p=1/4$	$p=1/8$	$p=1/16$	$p=1/64$	$p=1/256$
$\exp(-0.25v)$	0.0571	0.0256[1]	0.0178	0.0158[1]	0.0154	0.0152	
$\exp(-0.5v)$	0.114	0.0510[1]	0.0354	0.0314[1]	0.0305	0.0302	
$\exp(-v)$	0.224	0.100[1]	0.0694	0.061[1]	0.060	0.059	
$\exp(-2v)$	0.422	0.190[1]	0.128	0.113[1]	0.109	0.108	
$\exp(-4v)$	0.697	0.315	0.199	0.168	0.161	0.159	
$\exp(-8v)$	0.903	0.421	0.231	0.173	0.158	0.153	0.152
$\exp(-16v)$	0.975	0.476	0.235	0.140	0.111	0.101	0.100
$\exp(-32v)$	0.994	0.494	0.244	0.124	0.0759	0.0577	0.0565
$2v\ K_1(2v)$	0.168	0.072[1]	0.0602	0.060[1]	0.0583		
$4v\ K_1(4v)$	0.468	0.21[1]	0.156	0.15[1]	0.149		

[1] Interpolated values.

that some values are approximations, obtained by a crude interpolation. In each case, the variance for the lowest value of p gives a good approximation of the variance in a complete line-survey.

When the correlation is decreasing as $\exp(-4v)$ or $2vK_1(2v)$ or still slower, the variance is not very much reduced by a change of p from 1/8 to lower values. In all cases, however, the variance is substantially reduced when we pass from $p = 1$ to $p = 1/2$; and to a gradually smaller degree when p decreases to 1/4 and 1/8.

Supplementing table 24, the following limiting values are found when $h \to 0$ in $\exp(-hv)$:

p	1	1/2	1/4	1/8	1/16	1/64
Variance per unit distance	375.5	168.3	117.0	104.2	101.0	100.0

Here the variances are expressed in per cent of the value corresponding to $p = 1/64$.

Meanwhile, it may be noticed for the cases entered in table 24 that the variance is at least twice as high when $p = 1$ as it is in the case $p = 1/2$.

Consider then $4n$ equidistant sample points on the periphery of a unit square (fig. 15 A). Let $p = 1/n$ denote the distance between the sample points. The variance of the mean of the $4n$ points has been computed for some cases of exponential correlation and for p of the form $2^{-\nu}$, see table 25. The values tabulated thus are

$$(1/4n)^2 \sum_i \sum_j \exp(-h|x_i - x_j|) \tag{6.11.1}$$

It may be noted in some cases that (1) increases when the number of sample points is augmented. This indicates that some gain in efficiency may be achieved by giving the sample points different weights when calculating the mean. Disregarding this possibility of increasing the precision, we find that almost no efficiency is gained when p is passing from a value in the interval $(1/h > p > 1/2h)$ to any smaller value.

To obtain some further illustrations, we consider the correlations

$$0.5 \exp(-0.2v) + 0.5 \exp(-2.5v) \tag{6.11.2}$$
$$0.4 \exp(-v) + 0.6 \exp(-5v) \tag{6.11.3}$$
$$0.4 \exp(-v) + 0.6 \exp(-12v) \tag{6.11.4}$$
$$0.4 \exp(-3v) + 0.6 \exp(-12v) \tag{6.11.5}$$

Formula (2) gives a fairly good graduation of some empiric correlations in table 2 (series 3 and 4) of Ch. 4. The corresponding observations refer to the distribution of land area. The expressions (3)—(5) are taken from Matérn (1947). While (3) and (4) pertain to the distribution of forest land, (5) was

Table 25. Variance of the mean of observations from $4n$ sample points, located equidistantly on the sides of a unit square. Distance between sample points: $p=1/n$. Covariance function: exp(-hv).

h	$n=1$ $p=1$	$n=2$ $p=1/2$	$n=4$ $p=1/4$	$n=8$ $p=1/8$	$n=16$ $p=1/16$	$n=32$ $p=1/32$	$n=64$ $p=1/64$
0.25	0.815	0.829	0.834				
0.5	0.677	0.695	0.701				
1	0.495	0.504	0.510	0.511			
2	0.332	0.303	0.300	0.299	0.299	0.299	0.299
4	0.260	0.173	0.150	0.145	0.144	0.144	0.144
8	0.250	0.130	0.0842	0.0712	0.0680	0.0672	0.0671
16	0.250	0.125	0.0649	0.0416	0.0347	0.0329	0.0325

obtained by smoothing a correlogram derived from volumes on sample plots. It might also represent the areal distribution of a particular site class. (This is seen by comparing the columns 7 and 8 of table 5 in Matérn 1947, p. 60.) The expressions (4) and (5) have been used in previous sections of the present chapter (6.8.8 and 6.9.2). It should be noticed that the kilometer is unit of length in (2)—(5).

The variances per tract of size 1.6×1.6 km with a varying number (n) of sample points are shown in table 26 for the covariances (2)—(5). Tracts of this size have been used in region III in the current Swedish survey. Since some of the values are obtained graphically, only two figures are given in the table.

The values in the column "$n = 64$" can represent the variances with sufficient accuracy in a complete survey of the contour of the tract. When n is passing from $n = 4$ to $n = 64$ a reduction of the variance worth mentioning appears in the case (5) only. Even in this case it seems not reasonable to go beyond $n = 16$.

So far the results seem to indicate that the extra information on areal distributions which can be obtained from observations made between sample plots has a rather limited value when the distance between the plots is of the

Table 26. Variance of the mean of observations from $4n$ sample points, located equidistantly on the sides of a 1.6×1.6 square. Distance between sample points: $p = 1.6/n$. Four different covariance functions of exponential type.

Covariance function	$n=1$ $p=1.6$	$n=4$ $p=0.4$	$n=8$ $p=0.2$	$n=16$ $p=0.1$	$n=64$ $p=0.025$
(6.11.2)......................	0.52	0.48	0.47	0.47	0.47
(6.11.3)......................	0.30	0.19	0.19	0.18	0.18
(6.11.4)......................	0.30	0.18	0.17	0.16	0.16
(6.11.5)......................	0.25	0.087	0.068	0.063	0.061

order of 100—400 meters. This conclusion, however, should be regarded as tentative. More empirical evidence on the relevant topographic variation as well as economic data on the field work and office work connected with the different forms of survey, are required, if more definite recommendations are to be reached.

6.12. The size and shape of sample plots

In a paper (1938) Fairfield Smith examined a large number of published uniformity trials, chiefly from agricultural and horticultural fields. Studying the relationship between the variance per plot (V) and the size (A) of the plot, he found that most data could be described fairly well by a formula of the type

$$V = \text{const. } A^{-b} \tag{6.12.1}$$

The exponent b varied from field to field; most values belonged to the interval 0.2—0.8. (For methods of determining b empirically, see Hatheway & Williams 1958.) The shape of the plot did not seem to have any noticeable effect on V.

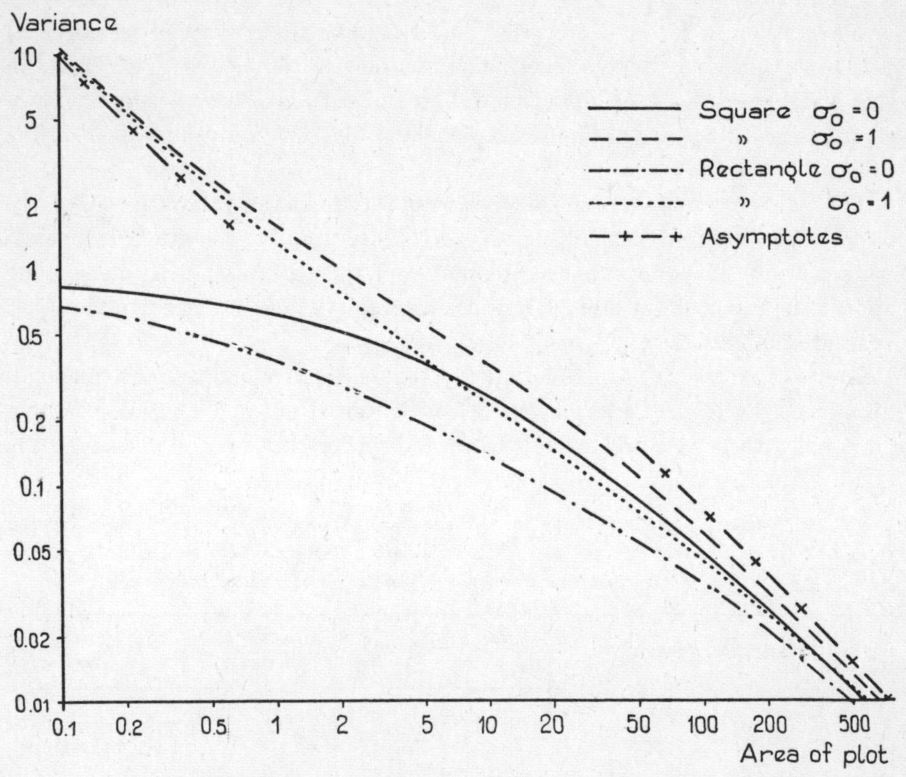

Fig. 16. Variance per unit area according to (6.12.2) with $c(v) = \exp(-v)$.

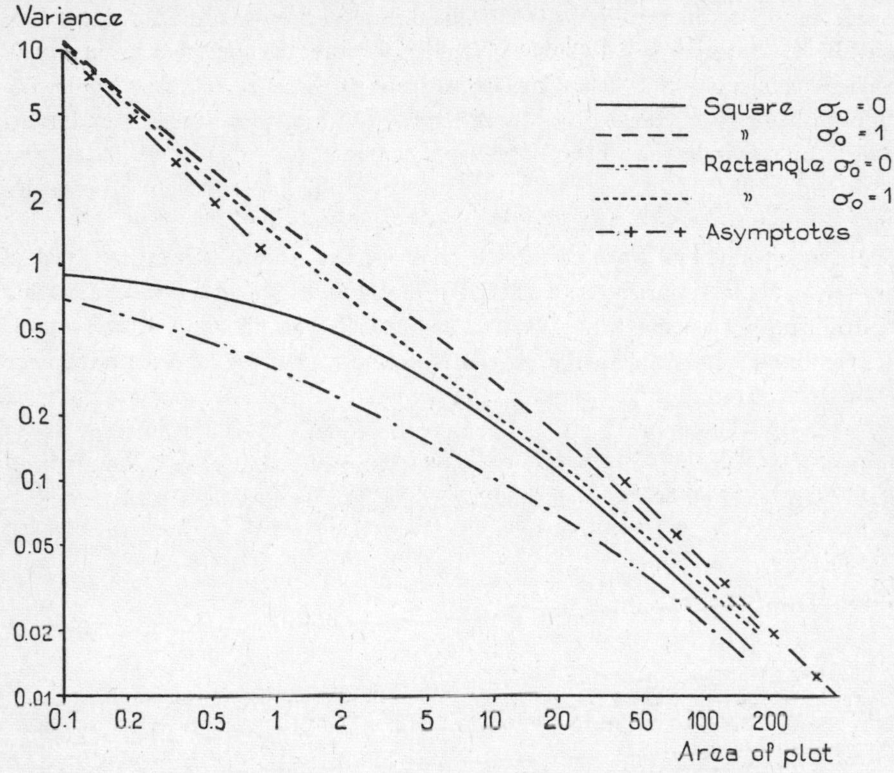

Fig. 17. Variance per unit area according to (6.12.2) with $c(v) = 2vK_1(2v)$.

The object of this section is to show that (1) can be obtained as a good approximation by considering an isotropic process.

We start from the following expression for the variance of an isotropic set function (cf. § 2.6)

$$V = D^2[z(q)] = \sigma_0^2/A + \int_0^\infty c(v)\, f(v;q)\, dv \tag{6.12.2}$$

Here A is the area of q; f is the frequency function of the distance between two random points in q; $c(v)$ is the cov.f. of the continuous component of z. The term σ_0^2/A should be conceived of as representing e.g. the influence of the random variation in plant number and the effect of errors of measurement. [It is then assumed that the observations are made for some small basic cells, and that $z(q)$ is an average of the data obtained from the basic cells in q.]

The following two covariance functions have been used

$$exp(-v) \tag{6.12.3}$$

$$2vK_1(2v) \tag{6.12.4}$$

Variances according to (2) have been computed for *squares* and for *oblong rectangles* (ratio 1 : 16 between the lengths of the sides). The area has been varied from 0.1 to 200—500. The expressions (3) and (4) and the two sets of figures have been combined with the two values $\sigma_0 = 0$ and $\sigma_0 = 1$ in (2). The basic computations have been described in § 5.2, where they have been utilized for other purposes. For this reason, a graphic presentation of the results seems to be sufficient here.

If logarithmic scales are used on both axes, the relationship (1) between V and A would be represented as a straight line. Fig. 16 shows the corresponding relationship for the cov.f. (3); case (4) is illustrated in fig. 17.

It is immediately apparent that the equation (2) can be represented closely enough by (1) over a fairly large range of A-values, if an appropriate value of σ_0 is chosen. The approximation seems to be equally good for both functions (3) and (4). The asymptotic properties of (2) indicate that similar results may hold for a wide class of correlation functions. For small values of A, (2) gives

$$V = \sigma_0^2/A + c(0) + \ldots \tag{6.12.5}$$

We then define

$$m_k = \int_0^\infty v^k\, c(v)\, dv$$

and assume that m_2 is finite. Using (2.5.19) we obtain the following development, valid for large A

$$V = \frac{\sigma_0^2 + 2\pi m_1}{A} - \frac{2 P m_2}{A^2} + \cdots \tag{6.12.6}$$

Here P denotes the perimeter of q. The asymptotic straight lines corresponding to (5) and (6) have been indicated in the figures. If a cov.f. is decreasing and has a finite m_2, the relationship between V and A must be roughly similar to those obtained with (3) and (4): an S-shaped curve traversing the band between (5) and (6) when A passes from 0 to ∞. Thus (1) can be expected to give a good approximation in a large number of cases. The problem of finding a cov.f. such that (1) is exactly valid for some range of A presents intricate difficulties, see Whittle (1956).

The figures do not corroborate Smith's observation that the variance should be largely independent of the *shape* of the plot. However, the asymptotes depend on the area A only. The dependence on shape is therefore weak for very small and very large values of A. In these cases V varies approximately as A^{-1}, whereas the exponents reported by Smith (1938) mostly were numerically smaller than 1, see above. It should be added in this context that many authors have given further evidence supporting "Fairfield Smith's law" (1) (for examples from forestry see Strand 1957). However, contrary to Smith (1938), several investigators have found a clear relationship between precision and shape of plot, i.a. Christidis (1931), Justesen (1932), Kalamkar (1932), and Bormann (1953).

Sammanfattning

Stokastiska modeller och deras tillämpning på några problem i skogstaxering och andra samplingundersökningar

Uppsatsen behandlar vissa statistiska problem, som inte kan utredas noggrant utan matematiska termer och symboler. I denna sammanfattning skall emellertid ett sådant framställningssätt inte nyttjas. Det blir därför här lika mycket fråga om en i huvudsak icke-matematisk kommentar till uppsatsen som om en sammanfattning.

Kap. 1. Inledning

Som överrubrik till uppsatsen har valts »Spatial variation». Någon god motsvarighet på svenska till denna term har förf. icke funnit. Vad för slags variation, som avses, kanske dock framgår av följande exempel:

lokaliseringen av mikroskopiska partiklar uppslammade i en vätska,

fördelningen av vintergator i rymden,

de geometriska mönster, som återfinnes på kartor över t. ex. bergarternas utbredning i en region,

det sätt på vilket individ av någon växt- eller djurart är utspridda över en lokal,

det mönster som bildas av ojämnheter på ytan av en fabricerad artikel av metall, trä, papper o. s. v.

För en mera begränsad klass av fenomen har förf. tidigare använt termen »topografisk variation». Denna beteckning avser dock inte enbart växlingar i topografin i egentlig mening utan syftar även på variation i fråga om markens bördighet, vegetationen, de klimatiska och geologiska faktorerna, o. s. v.

Den variation över en yta eller i en rymd varom det är fråga här, är i allmänhet så komplicerad och oregelbunden att man måste inskränka sig till en statistisk beskrivning. Detsamma gäller den variation man möter hos många slag av *tidsserier* (meteorologiska, ekonomiska o. s. v.). I teorin för *stokastiska processer*, som intar en framskjuten plats i modern matematisk statistik, har skapats en mångfald modeller eller schemata för sådana serier. Denna teori har efterhand generaliserats. Den kan nu sägas innefatta modeller för fenomen som variera i en godtycklig matematisk rymd. Utmärkande för dessa modeller är att de innehåller stokastiska (slumpmässiga) moment som enligt en eller annan mekanism sprider sitt inflytande i rummet eller tiden. Även om man sällan kommer fram till en fullt realistisk modell, kan man ofta finna goda approximationer. Teorin ger därtill begrepp och termer som är lämpliga för en allmän klassificering och en överslagsmässig beskrivning av olika företeelser.

De i denna uppsats redovisade undersökningarna sammanhänger med statistiska problem som möter vid planläggningen av en stickprovsundersökning av typ skogtaxering eller vid uppgörandet av fältförsöksplaner. Vid behandlingen av dessa problem är det väsentligt att man har en i stora drag riktig föreställning om strukturen hos variationen i det område undersökningen avser.

Kap. 2. Stationära stokastiska processer i R_n

Kapitlet behandlar matematiska modeller av fenomen, som varierar i 1, 2, 3 eller flera dimensioner, varvid variationen förutsättes vara stationär (homogen). Mycket vagt uttryckt innebär detta att variationen är av väsentligen samma struktur i alla delar av rymden.

Till varje punkt x antages höra ett värde $z(x)$, som anger tillståndet i punkten x hos det fenomen som studeras (t. ex. höjden över havet i punkten x). De enklaste egenskaperna hos en matematisk modell för variationen hos $z(x)$ kan uttryckas i medelvärdet (matematisk förväntan) och variansen för det enskilda värdet $z(x)$ och i korrelationen mellan värden $z(x)$ och $z(y)$, anknutna till två olika punkter. Antagandet om stationaritet betyder att medelvärdet och variansen är oberoende av x samt att korrelationen endast beror på det inbördes läget av punkterna x och y. Om korrelationen endast beror på avståndet är det fråga om en *isotropisk* modell.

Kap. 2 sysslar framför allt med egenskaperna hos de »korrelationsfunktioner» som uttrycker korrelationens beroende av två punkters inbördes läge. Ett stort antal exempel på isotropiska korrelationsfunktioner lämnas.

Genom en integration av till punkter knutna värden $z(x)$ kan man bilda summor och medeltal som avser områden (t. ex. genomsnittlig höjd över havet inom ett område). Allmänna formler meddelas för varianser och korrelationer som avser dylika medeltal.

Kap. 3. Några speciella modeller

Kapitel 3 ger en illustration till den föregående allmänna framställningen genom en tämligen detaljerad beskrivning av några speciella modeller. Det kan kanske här räcka med att peka på några mycket enkla exempel återgivna i fig. 1—3, sid. 49,50. Fig. 1 har erhållits genom en grov mekanism för en slumpmässig uppdelning av en plan region i två typer av områden. Fig. 2 och 3 visar »övernormal» resp. »undernormal» utspridning av punkter över ett plant område.

Kap. 4. Några anmärkningar om den topografiska variationen

Den empiriska motsvarigheten till en korrelationsfunktion är ett korrelogram, som anger hur graden av samvariation i ett område beror på avstånd och riktning. Man kan vänta sig att samvariationen skall vara stark mellan närbelägna punkter medan den bör förtunnas med växande avstånd. Detta har styrkts genom många undersökningar. I kap. 4 lämnas ytterligare bekräftelse genom ett par korrelogram.

I kapitlet diskuteras även hur korrelogrammet påverkas av observationsfel och avrundningsfel samt effekten av »lokal integration» och konkurrens om ljus, näring etc. (Exv. kan en plantas tillväxt sägas »integrera» bördighet och övriga betingelser i omgivningen.) En avslutande paragraf behandlar frågan om eventuell periodicitet i den topografiska variationen. Om en stark periodicitet föreligger, skulle en systematisk sampling (t. ex. ett regelbundet förband av ytor eller linjer) kunna giva helt vilseledande resultat. Ett av Finney (1950) anfört exempel på periodisk variation i resultat från en linjetaxering synes dock ej vara övertygande, vilket bl. a. utretts av Milne (1959).

Kap. 5. Om effektiviteten hos några metoder att välja samplingpunkter i planet

Man önskar uppskatta medeltalet av alla de värden en funktion $z(x)$ antar i ett plant område, Q. Som uppskattning av detta medeltal tages medeltalet, $\bar{z}$, av de z-värden som observeras i ett ändligt antal samplingpunkter i Q. Uppskattningens precision beror på hur funktionen varierar inom Q och på det sätt på vilket samplingpunkterna placeras ut. En diskussion av precisionen hos $\bar{z}$ kan vara vägledande även för lokalisering av andra typer av samplingenheter, t. ex. provytor och »trakter».

Det antas att z-värdena kan betraktas som resultatet av en viss slumpmässig procedur (z-värdena representerar med andra ord en realisation av en stokastisk process). Det förutsättes vidare att denna mekanism karakteriseras av en isotropisk och avtagande korrelationsfunktion. I kapitlet studeras vissa allmänna typer av sådana mekanismer. För detta slag av funktioner $z(x)$ göres en del jämförelser mellan olika metoder att lokalisera ett på förhand fixerat antal samplingpunkter i området Q.

Av de undersökta metoderna visar sig den systematiska samplingen i form av ett regelbundet triangelförband ge högst precision. Det regelbundna kvadratförbandet ger emellertid endast obetydligt större medelfel. Samplingfelen är således i dessa fall mindre än i de undersökta fallen av slumpmässig (»random») sampling. Vad beträffar s. k. stratifierad sampling, ger strata i form av regelbundna sexhörningar den högsta precisionen. En viss förbättring kan åstadkommas genom »djup stratifiering» t. ex. med anordningar av typ »romersk kvadrat». Dessa jämförelser är av intresse i de fall då samplingkostnaden huvudsakligen beror på antalet samplingenheter och inte på längden av den väg som man måste följa för att uppsöka samplingpunkterna (ex. sampling från karta). Några experimentella samplingundersökningar har givit resultat i någorlunda överensstämmelse med dem som erhållits med de teoretiska modellerna. För att på rent empirisk väg få säkra hållpunkter för ett val av samplingmetodik erfordras emellertid utomordentligt omfattande undersökningar.

Vid sampling i fält är kostnaden rätt mycket avhängig av den vägsträcka man måste tillryggalägga för att uppsöka alla samplingpunkter. Kap. 5 innehåller några data om väglängder (fågelvägen) för olika former av punktsampling. Medan dessa längder omedelbart kan anges i fråga om sampling med regelbundna förband av punkter, fordras det rätt ingående geometriska överväganden för att man skall kunna uppskatta motsvarande genomsnittliga längder vid de olika formerna av slumpmässig sampling. Om kostnaden beror *endast* på väglängden visar det sig att det bästa av de undersökta systemen är ett rektangelförband, där punkterna ligger oändligt tätt i ena riktningen, d. v. s. ett system av parallella och ekvidistanta linjer.

Kap. 6. Diverse problem sammanhörande med samplingundersökningar

Den inledande paragrafen till kap. 6 avser att ge en bakgrund till de utredningar som redovisas i kapitlet. Då en del av dessa undersökningar knyter an till den nu pågående svenska riksskogstaxeringen, lämnas även uppgifter om den samplingmetodik som tillämpas vid denna taxering.

I § 6.2 diskuteras sampling av en endimensionell population (tänkbara tillämpningar: sampling av en tidsserie, skogstaxering med parallella taxeringsbälten). Det systematiska stickprovet är, som påvisats av många författare, överlägset de slumpmässiga metoderna för vissa typer av populationer. I paragrafen behandlas även de risker för snedvridning av den systematiska samplingens resultat som kan uppkomma om en samplad tidsserie innehåller en mer eller mindre utpräglat periodisk komponent.

Ibland stöter man på principiella svårigheter när det gäller att bedöma ett stickprovs noggrannhet med hjälp av det insamlade siffermaterialet. Ett sådant fall behandlas i § 6.3, nämligen stratifierad sampling med *en* samplingenhet per stratum.

Vid vissa tillämpningar av »dubbel sampling» är det särskilt viktigt att man förfogar över korrekta metoder för skattning av precisionen. Ett exempel är följande. För ett stort sampel av provytor görs en okulär bestämning av t. ex. stamvolymen enligt ögonmått (i fält eller på flygbild). För ett delstickprov görs en noggrann uppmätning, vilken användes till kalibrering av den okulära bedömningens värden. I detta fall kommer valet av metod för precisionsuppskattningen att direkt påverka även volymsuppskattningen. Som antydes genom ett exempel i § 6.4 kan en olämplig medelfelsformel medföra att det kombinerade stickprovet utnyttjas mycket dåligt. Skattningen av volymen kan bli sämre än den man skulle få av enbart delstickprovets mätningar.

I paragraferna 6.5 och 6.6 behandlas en del metoder för uppskattning av medelfelet till systematiska stickprov på en linje resp. i ett plan. Dessa paragrafer, liksom ett tidigare arbete av förf. (1947), avser fallet med ett »stort sampel». Till komplettering lämnas i § 6.7 anvisningar för det fall att stickprovet är så litet att man måste taga hänsyn till kanteffekter.

I nästa avsnitt, 6.8, diskuteras »trakt-sampling», varvid med trakt avses en sluten plan kontur (cirkel, kvadrat, rektangel o. s. v.). För den tidigare antydda klassen av stokastiska processer har en del beräkningar utförts rörande precisionen hos olika slag av trakter, se fig. 14, sid. 124. Det visar sig att av de undersökta slutna figurerna med given omkrets ger cirkeln den högsta precisionen. Kvadraten är i sin tur överlägsen en avlång rektangel med samma perimeter. Om man vidgar jämförelsen till att avse även andra kontinuerliga kurvor än de slutna, synes det räta linjesegmentet (fig. 14, a) vara bättre än andra kurvor med samma längd. (De beräkningar som avser cirkeln torde kunna ge en viss föreställning om precisionen hos regelbundna sex- och åttahörningar.)

I § 6.9 jämföres två olika metoder att placera ut provytor på sidorna av en kvadrat, se fig. 15, (A) och (B), sid. 126. Jämförelsen ger ett svagt men klart företräde åt system (A). En något högre precision erhålles med system (C), där de sexton ytorna är placerade längs en åttasiding med samma omkrets som kvadraterna i (A) och (B).

Därefter följer, i § 6.10, en diskussion av frågan hur storleken av en trakt påverkar samplingfelen. Framställningen knyter an till skattningar grundade på provytor placerade på samma sätt som »stubb-ytorna» vid den tredje svenska riksskogstaxeringen, d. v. s. med 100 meters mellanrum längs periferin av en kvadrat. Beräkningar har utförts för kvadrater med sidolängder varierande från 1 200 till 2 200 meter. Om trakterna är utplacerade genom ett rent slumpmässigt förfarande (»unrestricted random sampling»), synes man kunna räkna med en varians (per trakt) som approximativt är omvänt proportionell mot traktsidans längd. Beräkningarna är grundade på korrelationsfunktioner som erhållits vid utjämning av korrelogram från skogstaxeringsmaterial.

Vissa observationer vid skogstaxering gäller arealens fördelning: ägoslag, bonitet o. s. v. Man kan ifrågasätta om en kontinuerlig registrering av sådana observationer längs taxeringslinjer eller traktsidor är nämnvärt bättre än en registrering som inskränkes till intermittenta punkter på linjerna eller traktsidorna. Vissa beräkningar har utförts på grundval bl. a. av grafiskt utjämnade korrelogram (§ 6.11).

De tycks ge vid handen att man (med den stickprovstäthet man har vid de svenska riksskogstaxeringarna) torde vinna föga genom att observationer över arealfördelningen läggs på kortare inbördes avstånd än något hundratal meter.

Ett viktigt problem vid skogstaxering och vid sampling av åkerjord etc. är vilken form och storlek de undersökta provytorna skall ha. Empiriska undersökningar har visat att sambandet mellan variansen och provytans storlek ungefärligen följer en formel, som brukar benämnas »Fairfield Smith's lag». I den avslutande paragrafen, 6.12, visas att man erhåller resultat i ganska god överensstämmelse med denna lag även då observationerna avser matematiska modeller av det slag som behandlats i detta arbetes tidigare avsnitt.

References

BARNES, R. M., 1957: Work sampling. Second edition. — New York.

BARTLETT, M. S., 1954: Processus stochastiques ponctuels. — Ann. Inst. Poincaré, 14:35—60.

— 1955: An introduction to stochastic processes. — Cambridge.

BAUERSACHS, E., 1942: Bestandesmassenaufnahme nach dem Mittelstammverfahren des zweitkleinsten Stammabstandes. — Forstwiss. Centralblatt, 64:182—186.

BITTERLICH, W., 1956: Die Relaskopmessung in ihrer Bedeutung für die Forstwirtschaft. — Österr. Vierteljahresschr. für Forstwesen, 97: 86—98.

BLANC—LAPIERRE, A. & FORTET, R., 1953: Théorie des fonctions aléatoires. — Paris.

BLOCK, E., 1948: Undersökningar över follikelapparatens variationer. — Lund.

BOCHNER, S., 1932: Vorlesungen über Fouriersche Integrale. — Leipzig.

BOREL, É. & LAGRANGE, R., 1925: Principes et formules classiques du calcul des probabilités. — Paris.

BORMANN, F. H., 1953: The statistical efficiency of sample plot size and shape in forest ecology. — Ecology, 34: 474—487.

BUCKLAND, W. R., 1951: A review of the literature of systematic sampling. — Jour. Roy. Stat. Soc., B 13: 208—215.

CARSTEN, H. R. F. & MCKERROW, N. W., 1944: The tabulation of some Bessel functions $K_\nu(x)$ and $K'_\nu(x)$ of fractional order. — Phil. Mag., 35, 7th series: 812—818.

CESARI, L., 1956: Surface area. — Princeton.

CHRISTIDIS, B. G., 1931: The importance of the shape of plots in field experimentation. — Jour. Agric. Sci., 21: 14—37.

COCHRAN, W. G., 1946: Relative accuracy of systematic and stratified random samples for a certain class of populations. — Ann. Math. Stat., 17: 164—177.

— 1953: Sampling techniques. — New York.

COTTAM, G. & CURTIS, J. T., 1956: The use of distance measures in phytosociological sampling. — Ecology, 37: 451—460.

CRAMÉR, H., 1937: Random variables and probability distributions. — Cambridge.

— 1940: On the theory of stationary random processes. — Ann. Math., 41: 215—230.

— 1945: Mathematical methods of statistics. — Uppsala.

DALENIUS, T., 1957: Sampling in Sweden. — Stockholm.

DAS, A. C., 1950: Two dimensional systematic sampling and the associated stratified and random sampling. — Sankhyā, 10: 95—108.

DELTHEIL, R., 1926: Probabilités géométriques. — Paris.

DOOB, J. L., 1953: Stochastic processes. — New York.

DURBIN, J., 1958: Sampling theory for estimates based on fewer individuals than the number selected. — Bull. Inst. Int. Stat. 36, 3: 113—119.

FAURE, P., 1957: Sur quelques résultats relatifs aux fonctions aléatoires stationnaires isotropes introduites dans l'étude expérimentale de certains phénomènes de fluctuations. — Comptes rendus, Paris, 244: 842—844.

FEJES, L., 1940: Über einen geometrischen Satz. — Math. Zeitschr., 46: 83—85.

FEJES TÓTH, L., 1953: Lagerungen in der Ebene auf der Kugel und im Raum. — Berlin.

FELLER, W., 1943: On a general class of "contagious" distributions. — Ann. Math. Stat., 14: 389—400.

— 1950: An introduction to probability theory and its applications, I. — New York.

FEW, L., 1955: The shortest path and the shortest road through n points. — Mathematica, 2: 141—144.

FINNEY, D. J., 1948: Random and systematic sampling in timber surveys. — Forestry, 22: 64—99.

— 1950: An example of periodic variation in forest sampling. — Forestry, 23: 96—111.

FOX, M., 1958: Effect of expansion of the universe on the distribution of images of galaxies on photographic plates. A simplified model. — Astron. Jour., 63: 266—272.
GHOSH, B., 1943: On the distribution of random distances in a rectangle. — Science and Culture, 8: 388.
— 1949: Topographic variation in statistical fields. — Calcutta Stat. Ass. Bull., 2: 11—28.
— 1951: Random distances within a rectangle and between two rectangles. — Bull. Calcutta Math. Soc., 43: 17—24.
GHOSH, M. N., 1949: Expected travel among random points in a region. — Calcutta Stat. Ass. Bull., 2: 83—87.
GRAB, E. L. & SAVAGE, I. R., 1954: Tables of the expected value of $1/X$ for positive Bernoulli and Poisson variables. — Jour. Am. Stat. Ass., 49: 169—177.
GRENANDER, U. & ROSENBLATT, M., 1956: Statistical analysis of stationary time series. — Stockholm.
GRENANDER, U. & SZEGÖ, G., 1958: Toeplitz forms and their applications. — Berkeley.
GURLAND, J., 1958: A generalized class of contagious distributions. — Biometrics, 14: 229—249.
HAGBERG, E., 1957: The new Swedish national forest survey. — Unasylva, 11: 3—8,28.
— 1958: Skogsuppskattningsmetoder vid skogsvärdering. — Sv. Lantmäteritidskr., 50: 330—337.
HÁJEK, J., 1959: Optimum strategy and other problems in probability sampling. — Časopis pro pěstováni matematiky, 84: 387—423.
HAMAKER, H. C., 1958: On hemacytometer counts. — Biometrics, 14: 558—559.
HAMMERSLEY, J. M., 1950: The distribution of distance in a hypersphere. — Ann. Math. Stat., 21: 447—452.
HAMMERSLEY, J. M. & NELDER, J. A., 1955: Sampling from an isotropic gaussian process. — Proc. Cambr. Phil. Soc., 51: 652—662.
HANSEN, M. H., HURWITZ, W. N. & MADOW, W. G., 1953: Sample survey methods and theory, I—II. — New York.
HARRIS, J. A., 1915: On a criterion of substratum homogeneity (or heterogeneity) in field experiments. — The Am. Naturalist, 49: 430—454.
HARTMAN, Ph. & WINTNER, A., 1940: On the spherical approach to the normal distribution law. — Am. Jour. of Math., 62: 759—779.
HATHEWAY, W. H. & WILLIAMS, E. J., 1958: Efficient estimation of the relationship between plot size and the variability of crop yields. — Biometrics, 14: 207—222.
HUDSON, H. G., 1941: Population studies with wheat. II, Propinquity. — Jour. Agric. Sci., 31: 116—137.
HUSU, A. P., 1957: O nekotorych funkcionalach na slučajnych poljach. (Summary: On some functionals on random fields.) — Vestnik, Leningrad. Univ., 1: 37—45, 208.
ITÔ, K., 1954: Stationary random distributions. — Mem. Coll. Sci. Univ. Kyoto, A 28: 209—223.
JAGLOM, *see* Yaglom.
JAHNKE, E. & EMDE, F., 1945: Tables of functions. Fourth ed. — New York.
JOHNSON, F. A. & HIXON, H. J., 1952: The most efficient size and shape of plot to use for cruising in old-growth Douglas-fir timber. — Jour. Forestry, 50: 17—20.
JOWETT, G. H., 1955: Sampling properties of local statistics in stationary stochastic series. — Biometrika, 42: 160—169.
JUSTESEN, S. H., 1932: Influence of size and shape of plots on the precision of field experiments with potatoes. — Jour. Agric. Sci., 22: 366—372.
KALAMKAR, R. J., 1932: Experimental error and the field-plot technique with potatoes. — Jour. Agric. Sci., 22: 373—385.
KARHUNEN, K., 1947: Über lineare Methoden in der Wahrscheinlichkeitsrechnung. — Ann. Acad. Sci. Fenn., A I: 37.
— 1952: Über ein Extrapolationsproblem in dem Hilbertschen Raum. — Compte rendu du onziéme Congrés de mathématiciens scandinaves tenu a Trondheim le 22—25 août 1949, pp. 35—41.
KENDALL, M. G., 1946: Contributions to the study of oscillatory time-series. — Cambridge.
KENDALL, M. G. & BUCKLAND, W. R., 1957: A dictionary of statistical terms. — Edinburgh & London.
KHINTCHINE, A., 1934: Korrelationstheorie der stationären stochastischen Prozesse. — Math. Annalen, 109: 604—615.

KILANDER, Kj., 1957: Några synpunkter på den skogliga tidsstudiemetodiken vid studium av virkestransporter. (Summary: Some viewpoints on the time study methods in forestry as applied to the study of timber transports.) — SDA-Redogörelse av intern natur, nr 5, 1957. (Mimeogr.)

LADELL, J. L., 1959: A method of measuring the amount and distribution of cell wall material in transverse microscope sections of wood. — Jour. Inst. Wood Sci., 3: 43—46.

LANGSAETER, A., 1926: Om beregning av middelfeilen ved regelmessige linjetakseringer. (Summary: Über die Berechnung des Mittelfehlers des Resultates einer regelmässigen Linientaxierung.) — Medd. fra det norske Skogforsøksvesen, 2 h. 7: 5—47.

— 1932: Nøiaktigheten ved linjetaksering av skog, I. (Summary: Accuracy in strip survey of forests, I.) — Medd. fra det norske Skogsforsøksvesen, 4: 431—563.

LHOSTE, E., 1925: Mém. de l'artillerie française 1925, pp. 245, 1027.

LOÈVE, M., 1948: Fonctions aléatoires du second ordre. (Note à P. Lévy: Processus stochastiques et mouvement brownien, pp. 299—352.) — Paris.

LORD, R. D., 1954: The use of the Hankel transform in statistics, I, II. — Biometrika, 41: 44—55, 344—350.

— 1954 a: The distribution of distance in a hypersphere. — Ann. Math. Stat., 25: 794—798.

MADOW, W. G. & L. H., 1944: On the theory of systematic sampling, I. — Ann. Math. Stat., 15: 1—24.

MAHALANOBIS, P. C., 1944: On large-scale sample surveys. — Phil. Trans. Roy. Soc., B 231: 329—451.

MARKS, E. S., 1948: A lower bound for the expected travel among *m* random points. — Ann. Math. Stat., 19: 419—422.

MATÉRN, B., 1947: Metoder att uppskatta noggrannheten vid linje- och provytetaxering. (Summary: Methods of estimating the accuracy of line and sample plot surveys.) — Medd. från statens skogsforskningsinst., 36, nr 1.

— 1953: Sampling methods in forest surveys. — Report on the seminar on advanced sampling, Stockholm, Nov. 1952, pp. 36—48. FAO, Rome. (Mimeogr.)

— 1959: Några tillämpningar av teorin för geometriska sannolikheter. (Summary: Some applications of the theory of geometric probabilities.) — Sv. Skogsvårdsfören. Tidskr., 57: 453—458.

MILNE, A., 1959: The centric systematic area-sample treated as a random sample. — Biometrics, 15: 270—297.

NAIR, K. R., 1944: Calculation of standard errors and tests of significance of different types of treatment comparisons in split-plot and strip arrangements of field experiments. — Indian Jour. Agric. Sci., 14: 315—319.

NÄSLUND, M., 1939: Om medelfelets härledning vid linje- och provytetaxering. (Summary: On computing the standard error in line and sample plot surveying.) — Medd. från statens skogsförsöksanstalt, 31: 301—344.

NEYMAN, J., 1939: On a new class of "contagious" distributions, applicable in entomology and bacteriology. — Ann. Math. Stat., 10: 35—57.

NEYMAN, J. & SCOTT. E. L., 1958: Statistical approach to problems of cosmology. — Jour. Roy. Stat. Soc., B 20: 1—43.

ÖSTLIND, J., 1932: Erforderlig taxeringsprocent vid linjetaxering av skog. (Summary: The requisite survey percentage when line-surveying a forest.) — Sv. Skogsvårdsfören. Tidskr., 30: 417—514.

PATTERSON, H. D., 1954: The errors of lattice sampling. — Jour. Roy. Stat. Soc., B 16: 140—149.

PYKE, R., 1958: On renewal processes related to type I and type II counter models. — Ann. Math. Stat., 29: 737—754.

QUENOUILLE, M. H., 1949: Problems in plane sampling. — Ann. Math. Stat., 20: 355—375.

RYSHIK, I. M. & GRADSTEIN, I. S., 1957: Summen-, Produkt- und Integraltafeln. — Berlin.

SAKS, S., 1937: Theory of the integral. Second ed. — Warszawa.

SANTALÓ, L. A., 1953: Introduction to integral geometry. — Paris.

SAVELLI, M., 1957: Étude expérimentale du spectre de la transparance locale d'un film photographique uniformément impressionné. — Comptes rendus, Paris, 244: 871—873.

SCHOENBERG, I. J., 1938: Metric spaces and completely monotone functions. — Ann. Math., 39: 811—841.

SEGEBADEN, G. VON, 1964: Studies of cross-country transport distances and road net extension. — *Studia forestalia suecica*, 18.

SETH, S. K., 1955: Sampling and assessment of forest crops. — Forest Department, Uttar Pradesh, Bull. No. 21.

SIMONSEN, W., 1947: Om Transformation af Integraler af reelle Funktioner i abstrakte Rum. — Matem. Tidsskr. B, Aargang 1947: 55—61.

SKELLAM, J. G., 1952: Studies in statistical ecology, I. — Biometrika, 39: 346—362.

— 1958: On the derivation and applicability of Neyman's type A distribution. — Biometrika, 45: 32—36.

SLUTZKY, E., 1937: The summation of random causes as the source of cyclic processes. — Econometrica, 5: 105—146.

SMITH, H. F., 1938: An empirical law describing heterogeneity in the yields of agricultural crops. — Jour. Agric. Sci., 28: 1—23.

SOLOMON, H., 1953: Distribution of the measure of a random two-dimensional set. — Ann. Math. Stat., 24: 650—656.

SOMMERVILLE, D. M. Y., 1929: An introduction to the geometry of *n* dimensions. — London.

STEINHAUS, H., 1954: Length, shape and area. — Colloquium Mathematicum, 3: 1—13.

STRAND, L., 1951: Feilberegning i forbindelse med skogtakasjon. — Tidsskr. for skogbruk, 59: 195—209.

— 1954: Mål for fordelingen av individer over et område. (Summary: A measure of the distribution of individuals over a certain area.)— Medd. fra det norske skogforsøksvesen, 12: 191—207.

— 1957: Virkningen av flatestørrelsen på nøyaktigheten ved prøveflatetakster. (Summary: The effect of the plot size on the accuracy of forest surveys.) — Medd. fra det norske skogforsøksvesen, 14: 621—633.

TEPPING, B. J., HURWITZ, W. N. & DEMING, W. E., 1943: On the efficiency of deep stratification in block sampling. — Jour. Am. Stat. Ass., 38: 93—100.

THOMAS, M., 1949: A generalization of Poisson's binomial limit for use in ecology. — Biometrika, 36: 18—25.

THOMPSON, H. R., 1954: A note on contagious distributions. — Biometrika, 41: 268—271.

— 1955: Spatial point processes, with application to ecology. — Biometrika, 42: 102—115.

THOMPSON, W. R., 1935: On a criterion for the rejection of observations and the distribution of the ratio of deviation to sample standard deviation. — Ann. Math. Stat., 6: 214—219.

THOMPSON, W. R. et al., 1932: The geometric properties of microscopic configurations, I, II. — Biometrika, 24: 21—38.

TIPPET, L. H. C., 1934: Statistical methods in textile research. — Shirley Inst. Mem., 13: 35—93.

TURNER, M. E. & EADIE, G. S., 1957: The distribution of red blood cells in the hemacytometer. — Biometrics, 13: 485—495.

VERBLUNSKY, S., 1951: On the shortest path through a number of points. — Proc. Amer. Math. Soc., 2: 904—913.

WATSON, G. N., 1944: A treatise on the theory of Bessel functions. Second ed. — Cambridge.

WHITTLE, P., 1954: On stationary processes in the plane. — Biometrika, 41: 434—449.

— 1956: On the variation of yield variance with plot size. — Biometrika, 43: 337—343.

WICKSELL, S. D., 1925, 1926: The corpuscle problem. — Biometrika, 17: 84—99; 18: 151—172.

WIDDER, D. V., 1941: The Laplace transform. — Princeton.

WILLIAMS, R. M., 1952: Experimental designs for serially correlated observations. — Biometrika, 39: 151—167.

— 1956: The variance of the mean of systematic samples. — Biometrika, 43: 137—148.

WINTNER, A., 1940: Spherical equidistributions and a statistics of polynomials which occur in the theory of perturbations. — Astron. papers dedicated to Elis Strömgren: 287—297. Khvn.

WOLD, H., 1934: Sheppard's correction formulae in several variables. — Skand. Aktuarietidskr., 17: 248—255.

— 1938: A study in the analysis of stationary time series. — Uppsala.

— 1948: Random normal deviates. — Cambridge.

— 1949: Sur les processus stationnaires ponctuels. — Actes du colloque de calcul de probabilités de Lyon (juin 1948), pp. 75—86.

YAGLOM, A. M., 1957: Nekotorye klassy slučajnych polej v n-mernom prostranstve, rodstvennye stacionarnym slučajnym processam. (Summary: Certain types of random fields in n-dimensional space similar to stationary stochastic processes.) — Teorija verojatnostej i ee primenenija, 2: 292—338.
— 1959: Einführung in die Theorie stationärer Zufallsfunktionen. — Berlin.
YATES, F., 1948: Systematic sampling. — Phil. Trans. Roy. Soc., A 241: 345—377.
— 1953: Sampling methods for censuses and surveys. Second ed. — London.
ZUBRZYCKI, S., 1957: O szacowaniu parametrów złóż geologicznych. (Summary: On estimating gangue parameters.) — Zastosowania Matematyki, 3: 105—153.
— 1958: Remarks on random, stratified and systematic sampling in a plane. —Colloquium Mathematicum, 6: 251—264.

Author Index

Subject Index

Postscript

This book was originally published in Reports of the Forest Research Institute of Sweden (Meddelanden från Statens Skogsforskningsinstitut), Vol. 49, No. 5 (1960).

In the present edition I have corrected only minor errors and misprints. An author index and a subject index have also been added. Due to the length of time since the first edition, a revision would have amounted almost to writing a new book. The reader interested in the present state of the art in spatial statistics may consult recent textbooks such as Ripley (1981), and Diggle (1983). The underlying theory of geometric probabilities and integral geometry can be studied, e.g., in Santaló (1976). As to applications, interest has (in recent years) focused on stereology, see, e.g., Weibel (1979, 1980). The works of the French geostatistical school also deserve particular mention, see Matheron (1971, 1975), Marbeau (1976), and Serra (1982).

I shall now give some comments on certain sections of the book.

The mathematical and probabilistic basis is sketched in Chapters 2 and 3. The presentation is restricted mainly to the second-order properties of spatial stochastic processes (random fields) which are sufficient for the applications which I have in mind. Somewhat to my surprise, the examples of mechanisms producing point patterns that are more regular than those of a two-dimensional Poisson process (Section 3·6)—which I considered slightly curious—have been much quoted by later authors. On the other hand, the roughly outlined stationary set functions in Section 2·6 do not seem to have received any attention, although the related class of random measures has been the subject of deep and extensive investigation, e.g., Kallenberg (1975) and Matheron (1975).

Chapter 4 treats empiric observation of planar variation. Sections 4·2 and 4·3 give data on second-order properties of variables observed in forest inventories. A large number of similar data (correlograms) from more recent forest surveys has been published by Ranneby (1981b). The observational phenomena studied in Sections 4·3 and 4·4 are analogous to the "nugget effect" and the "regularization" of geostatistics.

Chapters 5 and 6 are devoted to applications. In Chapter 5 the theory of random planar processes is applied to the comparison of various designs of plane sampling. In the discussion of two-dimensional systematic sampling (Section 5.4), I have assumed that border effects can be neglected. It was later pointed out that this assumption is not justified, see Tubilla (1975) and Ripley (1982). Ripley says, in discussing my results, that "some of the more extreme efficiency gains under stratified and systematic sampling are probably rather optimistic".

Section 5.5 gives *inter alia* some figures on the efficiency of estimating the area of a randomly located figure by different methods of point samples. It ought to be added that the systematic sample, i.e., the lattice of points, had been the subject of very close investigations by Kendall (1948) and Kendall & Rankin (1953). The oscillatory behaviour of the error variance as a function of the area of the figure – which manifests itself as certain irregularities in the figures of the last column of Table 10 – is very pronounced in the case of circles and ovals studied as by Kendall and Rankin. It has also been confirmed for rectangles, Matérn (1985). Subsequent studies have shown that this heavy oscillation is not a general rule: it is not found in the case of an equilateral triangle.

The remaining portions of Chapter 5 and Sections 6.8–6.12 contain studies of other designs used in areal sampling. Recent studies of this type (related to the Swedish National Forest Survey) have been reported by Ranneby (1979, 1981a). The performance of so-called point sampling (Bitterlich sampling) in "Poisson forests" and in realizations of some other planar point processes have been investigated by Holgate (1967) and Matérn (1972a).

As stated in Section 1.1, I had planned to make analogous studies of the efficiency of various designs of field experiments. Time did not permit any extensive investigations of this type; however, certain results are given in Matérn (1971, 1972b). They concern the expected error variance and the expected length of confidence intervals when a design is applied to realizations of stationary planar processes. A typical result is that the latin square has poorer efficiency than the randomised block for a field trial with a large number of treatments. In the paper (Matérn, 1972b), it is argued that a number of similar questions can be studied by the same means, such as the comparison of incomplete and complete blocks, the possible advantages of split-plot designs, systematic designs, the effect of guard rows, etc. The only later application of the random process theory to the study of field experiments, which I know about, is Duby *et al.* (1977).

Section 6.4 deals briefly with two-phase sampling in cases where the second phase is a systematic sample. It is shown that the estimation method is crucial. The problem may become highly important in combining satellite data with systematic field samples. It is analogous with the "co-kriging" in geostatistics.

To conclude this commentary, I think that I can maintain that the kind of applications treated here — to statistical problems in areal sampling and field experimentation — have not received very much attention recently. I hope that this reprint, which is addressed to a wider circle of statisticians than the original publication, can stimulate more interest in similar applications of the theory of spatial stochastic processes.

References

DIGGLE, P. J., 1983: Statistical Analysis of Spatial Point Patterns. — Academic Press London.

DUBY, C., GUYON, X. & PRUM, B., 1977: The precision of different experimental designs for a random field. — Bimetrika, 64: 59–66.

HOLGATE, P., 1967: The angle-count method. — Bimetrika, 54: 615–623.

KALLENBERG, O., 1975: Random Measures. — Akademie-Verlag, Berlin.

KENDALL, D. G., 1948: On the number of lattice points inside a random oval. — Quart. J. of Math. (Oxford), 19: 1–26.

KENDALL, D. G. & RANKIN, R. A., 1953: On the number of points of a given lattice in a random hypersphere. — Quart. J. of Math. (Oxford), 2nd series, 4: 178–189.

MARBEAU, J.-P., 1976: Géostatistique Forestière. — Thèse pour le doctorat, Ecole Nationale Supérieure des Mines de Paris.

MATHERON, G., 1971: The theory of regionalized variables and its applications. — Les Cahiers du Centre de Morphologie Mathématique de Fontainebleau, No. 5.

MATHERON, G., 1975: Random Sets and Integral Geometry. — Wiley, New York.

MATÉRN, B., 1971: Stochastic models of planar variation. (In Swedish, English summary). The third Nordic Conference on Mathematical Statistics, Umeå, 10th–13th June, 1969. Published by the Swedish Statistical Association.

MATÉRN, B., 1972a: The precision of basal area estimates. — Forest Science, 18: 123–125.

MATÉRN, B., 1972b: Performance of various designs of field experiments when applied in random fields. — Third Conference of the Advisory Group of Forest Statisticians. Jouy-en-Josas, 7th–11th Sept. 1970, 119–129. Institut National de la Recherche Agronomique.

MATÉRN, B., 1985: Estimating area by dot counts. — Contributions to Probability and Statistics in Honour of Gunnar Blom (J. Lanke & G. Lindgren, eds.), 243–257. Department of Mathematical Statistics, University of Lund.

RANNEBY, B., 1979: Model studies of tract sizes in forest survey. — Forest Resource Inventories (W. E. Frayer, ed.), Vol. 3, 289–297.

RANNEBY, B., 1981a: The importance of plot size in forest inventories. (In Swedish.) — Swedish Univ. of Agric. Sciences, Department of Forest Survey, Project NYTAX 83, Report No. 4.

RANNEBY, B., 1981b: The spatial variation of some forest variables. A presentation of estimated correlation functions. (In Swedish.) — Swedish Univ. of Agric. Sciences, Department of Forest Survey, Project NYTAX 83, Report No. 5.

RIPLEY, B. D., 1981: Spatial Statistics. — Wiley, New York.

RIPLEY, B. D., 1982: Edge effects in spatial stochastic processes. — Statistics in Theory and Practice (B. Ranneby, ed.). Swedish Univ. of Agric. Sciences, Section of Forest Biometry, 247–262.

SANTALÓ, L. A, 1976: Integral Geometry and Geometrical Probability. — Addison-Wesley, Reading, Mass.

SERRA, I., 1983: Image Analysis and Mathematical Morphology. — Academic Press, London.

TUBILLA, A., 1975: Error convergence rates for estimates of multidimensional integrals of random functions. — Stanford University, Department of Statistics, Technical Report No. 72.

WEIBEL, E. R., 1979: Stereological Methods. Vol. 1. Practical Methods for Biological Morphometry. — Academic Press, London.

WEIBEL, E. R., 1980: Stereological Methods. Vol. 2. Theoretical Foundations. — Academic Press, London.